Cinema, Audiences and N

The purpose of this book is to shed new light on the cinema and modernity debate by confronting established theories on the role of the modern cinematic experience with new empirical work on the history of the social experience of cinemagoing, film audiences and film exhibition in Europe.

The book provides a wide range of research methodologies and perspectives on these matters, including:

- the use of oral history methods
- questionnaires
- diaries
- audience letters
- industrial, sociological and other accounts on historical film audiences.

The collection's case studies thus provide a 'how to' compendium of current methodologies for researchers and students working on film and media audiences, film and media experiences, and historical reception.

The volume is part of a 'new cinema history' movement within film and screen studies to look at cinema history not only as a history of production, textual relations or movies-as-artefacts, but rather to concentrate on the reception and social experience of cinema, and the engagement of film/cinema (history) 'from below'. The contributions to the volume reflect on the very different ways in which cinema has been accepted, rejected or disciplined as an agent of modernity in neighbouring parts of Europe, and on how cinemagoing has been promoted and regulated as a popular social practice at different times in twentieth-century European history.

Daniel Biltereyst is Professor in Film and Media Studies at the Department of Communication Studies, Ghent University, Belgium, where he leads the Centre for Cinema and Media Studies (CIMS). His research on film and screen culture as sites of controversy and censorship has been published in *Cultural Policy*, *European Journal of Cultural Studies*, *Historical Journal of Film, Radio and Television*, *Journal of Communication Inquiry*, *Media, Culture & Society*, *Screen*, *Studies in French Cinema*, *Studies in Russian and Soviet Cinema*.

Richard Maltby is Professor of Screen Studies and Deputy Executive Dean of the Faculty of Education, Humanities, Law and Theology at Flinders University, South Australia. He is Series Editor of Exeter Studies in Film History, the author of over 50 articles and essays, and the lead investigator on two Australian Research Council Discovery projects examining the structure of the distribution and exhibition industry and the history of cinema audiences in Australia.

Philippe Meers is an Associate Professor in Film and Media Studies in the Department of Communication Studies at the University of Antwerp, Belgium. His publications on popular media culture and film audiences have appeared in *Screen Media, Culture & Society*, *The Journal of Popular Film and Television*, *Iluminace* and other journals. He was the lead investigator on The 'Enlightened' City-project on the history of film exhibition and film culture in Flanders and Brussels (2005–8, with Daniel Biltereyst and Marnix Beyen).

Cinema, Audiences and Modernity

New Perspectives on
European Cinema History

Edited by
Daniel Biltereyst, Richard Maltby and
Philippe Meers

Routledge
Taylor & Francis Group

LONDON AND NEW YORK

First published 2012
by Routledge
2 Park Square, Milton Park, Abingdon, Oxon OX14 4RN

Simultaneously published in the USA and Canada
by Routledge
711 Third Avenue, New York, NY 10017

Routledge is an imprint of the Taylor & Francis Group, an informa business

British Library Cataloguing in Publication Data
A catalogue record for this book is available from the British Library

Library of Congress Cataloging in Publication Data
Cinema audiences and modernity: an introduction / Daniel Biltereyst [editor].
p. cm.
Includes bibliographical references and index.
1. Motion picture audiences: Europe: History: 20th century. 2. Motion picture
industry: Europe: History: 20th century. I. Biltereyst, Daniël, 1962-
PN1995.9.A8C57 2011
384'.8094: dc22
     2011009219

ISBN: 978-0-415-67277-1 (hbk)
ISBN: 978-0-415-67278-8 (pbk)
ISBN: 978-0-203-80463-6 (ebk)

Typeset in Bembo
by Taylor & Francis Books

MIX
Paper from
responsible sources
FSC
www.fsc.org FSC® C004839

Printed and bound in Great Britain by the MPG Books Group

Contents

List of figures and tables

Figures

Tables

Contributors

Daniel Biltereyst is Professor of Film and Media Studies, head of the Department of Communication Studies and director of the Centre for Cinema and Media Studies at Ghent University, Belgium. He has published on screen culture, controversy and the public sphere in journals such as *Media, Culture & Society, Historical Journal of Film, Radio and Television* and *Screen,* and in edited collections, including recent contributions to *Going to the Movies* (2008), *Watching the Lord of the Rings* (2008), *Je t'aime … moi non plus: Franco-British Cinematic Relations* (2010), *Billy Wilder, Moviemaker* (2011), and *The Handbook of Political Economy of Communications* (2011). With Richard Maltby and Philippe Meers, he edited *Explorations in New Cinema History: Approaches and Case Studies* (2011). He is currently editing a volume on film censorship around the world (with Roel Van de Winkel).

Pierluigi Ercole completed his PhD at the University of East Anglia and has taught at the University of Sussex. His areas of research include Anglo-Italian relations and the film industry during the silent era, cinemagoing and diasporic culture, Italian cinema. His work has been published in *Laboratorio di Nuova Ricerca: Investigating Gender, Translation and Culture in Italian Studies.* Articles on the distribution of Italian films in Britain will be published in *Silent Italian Cinema: A Reader* and in *Film History.*

Andrea Haller is Curator at the German Filmmuseum in Frankfurt am Main. She studied German literature, social anthropology, sociology and film studies at the University of Trier, Germany. In 2009 she was awarded her PhD for a thesis on early cinema's programming strategies and the female audience. She has published on local film history, programming practices of early cinema, female moviegoing and fandom and early movie stars. Her most recent publication is 'Seen through the eyes of Simmel: The Cinema Program as a "Modern" Experience', in *Film 1900. Technology, Perception, Culture* (2009).

Åsa Jernudd holds a doctoral degree in Cinema Studies from Stockholm University and is a senior lecturer at Örebro University, Sweden. Her publications include 'Educational Cinema and Censorship in Sweden, 1911–21', in *Nordic Explorations: Films Before 1930* (1999). Other articles have been published in journals such as *Film History* and *Aura: Film Studies Journal*.

Frank Kessler is Professor of Film and Television Studies at Utrecht University, the Netherlands. He is a co-founder and co-editor of *KINtop. Jahrbuch zur Erforschung des frühen Films*. From 2003 to 2007 he was the president of *Domitor*, an international organization for research on early cinema. In 2009 he was a Fellow at the International Research Institute for Cultural Technologies and Media Philosophy in Weimar. Kessler's work has been published in journals such as *Versus, Montage/AV, Cinémas, Réseaux, Iris, 1895*, and is included in many anthologies, among which are *The German Cinema Book* (2002), *The Cinema of Attractions Reloaded* (2006), *Psyche im Kino. Sigmund Freud und der Film* (2006) and *Hollywood: Les fictions de l'exil* (2007). Recently Kessler published (with co-editor Nanna Verhoeff) *Networks of Entertainment: Early Film Distribution 1895–1915* (2007).

Sabine Lenk is a film archivist, and Affiliated Researcher at Utrecht University. From 2007–10 she headed the film collections department at the Dutch Institute for Sound and Vision. From 1999–2007, she was the director of the Filmmuseum Düsseldorf. She is a co-founder of *KINtop. Jahrbuch zur Erforschung des frühen Films*. She has published widely on film archiving, cinema museology and early cinema in journals such as *Film History, Montage/AV, 1895, Maske und Kothurn, Journal of Film Preservation* and *Archives*. Her work is also published in anthologies including *La firme Pathé frères 1896–1914* (2004), *Psyche im Kino. Sigmund Freud und der Film* (2006) and *Le 'local' dans l'histoire du cinéma* (2008). She has published the anthologies *Obsessionen. Die Alptraum-Fabrik des Alfred Hitchcock* (2000) and *Grüße aus Viktoria. Film-Ansichten aus der Ferne* (2002). Her most recent book is *Vom Tanzsaal zum Filmtheater. Eine Kinogeschichte Düsseldorfs* (2009).

Annemone Ligensa is working on the research project 'Industrialization of Perception' at the University of Siegen, Germany, and finishing her dissertation. She received her MA degree at the University of Cologne, Germany, in Theater, Film and Television Studies/ English/ Psychology, and has been a lecturer in Media Psychology and assistant manager of the ZfMK (Center for Media Studies, Cologne). She has published articles in literature, film and media studies on the subjects of stardom, psychoanalytic film theory, narratology, reception and media theory. She is the co-editor of *Film 1900: Technology, Perception, Culture* (2009).

Martin Loiperdinger wrote his dissertation on Leni Riefenstahl's 'Triumph of the Will' (Goethe-University, Frankfurt, 1985). He has been Professor of Media Studies at the University of Trier, Germany, since 1998 and was an Assistant Professor at the University of Munich and an assistant at the University of Kassel. From 1993 to

1997, Loiperdinger was Deputy Director of the Deutsches Filminstitut, Frankfurt am Main. Loiperdinger is co-founder of *Kintop: Jahrbuch zur Erforschung des frühen Films*.

Kathleen Lotze is a PhD student and member of the Visual Studies and Media Culture Research Group at the University of Antwerp, Belgium. Her PhD project *Antwerp Cinema City. A media-historical investigation into the post-war evolution of film exhibition and reception in the city of Antwerp (1945–2010) with focus on the Rex-group* is a follow-up work to *The Enlightened City* project on the history of film exhibition and film culture in Flanders, for which she had been working as a researcher. Since 1999 she has been working for several media(historical) research projects at universities in Germany, The Netherlands and Belgium. She has published articles on underground film practice in East Germany in the 1980s.

Richard Maltby is Professor of Screen Studies and Executive Dean of the Faculty of Education, Humanities and Law at Flinders University, South Australia. His publications include *Hollywood Cinema: Second Edition* (2003), *'Film Europe' and 'Film America': Cinema, Commerce and Cultural Exchange, 1925–1939*, which won the Prix Jean Mitry for cinema history in 2000, and five edited books on the history of movie audiences and exhibition history. He is Series Editor of *Exeter Studies in Film History*, and the author of over 50 articles and essays. He is a Fellow of the Australian Academy of the Humanities. His co-authored book, *The New Cinema History: A Guide*, will be published in 2011.

Anna Manchin completed her dissertation on Fables of Modernity: Entertainment Films and the Social Imaginary in Interwar Hungary in 2008 at Brown University (Providence, RI). In 2008–9 she was a Ray D. Wolfe post-doctoral fellow at the University of Toronto. She is currently working on a book project on Jewish identities in interwar popular culture.

Philippe Meers is an Associate Professor in Film and Media Studies at the University of Antwerp, Belgium, where he is deputy director of the Visual Studies and Media Culture Research Group. He has published on popular media culture and film audiences in *Media, Culture and Society, Iluminace, The Journal of Popular Film and Television* and *Screen,* and in edited collections, including *Big Brother International* (2004); *Hollywood Abroad. Audiences and Cultural Relations* (2004); *The Lord of the Rings: Popular Cinema in Global Culture* (2007); *Watching The Lord of the Rings* (2007); *The Contemporary Hollywood Reader* (2009), and *The Handbook of Political Economy of Communications* (2011). With Richard Maltby and Daniel Biltereyst, he edited *Explorations in New Cinema History. Approaches and Case Studies* (2011). His current research focuses on historical and diasporic cinema cultures.

Stefan Moitra is a film and labour historian at University College London. He currently works on a PhD project on the place of cinema in British and German working-class cultures after 1945. He has been a research assistant at Ruhr-University

Bochum and teaches film history at Carl von Ossietzky University Oldenburg. He has contributed to *Films that Work: Industrial Film and the Productivity of Media* (2009), to *Das Spiel mit dem Fußball: Interessen, Projektionen und Vereinnahmungen* (2007), and has co-edited a collected volume on German labour relations, *Stimmt die Chemie? Mitbestimmung und Sozialpolitik in der Geschichte des Bayer-Konzerns* (2007).

Pavel Skopal is a lecturer at the Department of Film and Audiovisual Culture, Masaryk University, Brno, Czech Republic. He has published articles on the history of film distribution, exhibition, reception and on the practices of the DVD market in several periodicals and anthologies. He has edited an anthology devoted to the local cinema history in Brno. He is currently coordinating the collective project 'Local History of Cinema: Brno and Cultural History of Cinema to 1945', editing a book on cinema industry in post-war Czechoslovakia (*Pre-Planned Cinema. Czechoslovak Film Industry, 1945–1960*), and launching a comparative project on cinema cultural policy in Czechoslovakia and the GDR (1949–70) supported by the Alexander von Humboldt Foundation.

Petr Szczepanik is an Associate Professor at the Department of Film and Audiovisual Culture, Masaryk University, Brno, and a researcher at the National Film Archive (NFA), Prague. He is an editor-in-chief of *Iluminace*, a film studies journal published by NFA. He has published essays on the coming of sound in the context of broader media culture, the history of industrial film, and on film's interrelations with other technical media. He has also edited or co-edited three books on history of film thought and on film historiography, including *Cinema All the Time: An Anthology of Czech Film Theory and Criticism, 1908–1939* (2008). His latest book is called *Canned Words. Coming of Sound Film and Czech Media Culture of the 1930s* (2009). His current research project focuses on the history of Barrandov studios in Prague and the globalization of film production.

Lies Van de Vijver has a Master in Arts History (Ghent University) and Film Studies (University of Antwerp). She is a researcher at Ghent University, Belgium, where she works on the history of film exhibition, programming and cinemagoing. After working on *The Enlightened City* project, she is now engaged in a follow-up project *Gent Kinemastad* and is preparing a doctoral thesis on the Flemish film exhibition scene.

Thunnis Van Oort is a lecturer in Film Studies in the Department of Media and Culture Studies, Utrecht University, the Netherlands. He is the author of *Film en het Moderne Leven in Limburg* (*Film and modern life in Limburg*). His work has also been published in journals such as *Film History*.

Acknowledgements

Since the work of Georg Simmel and Walter Benjamin, the issue of cinema's relation to modernity has been central to writing on film, and has exercised a powerful influence on theoretical understandings of the experience of cinema. This collection seeks to engage and confront the issue of cinema's modernity with new empirical work on the history of the social experience of cinemagoing, film audiences and film exhibition. It has its origins in a conference on new methodologies for historical film studies held in Ghent in December 2007 under the title, 'The Glow in their Eyes'. One of this event's recurrent topics was an engagement with the cinema/modernity-debate. A series of papers tackled this issue from such fresh and empirically grounded perspectives that we as conference organizers rapidly realized the value in bringing this new work together in a publication that would shed new light upon this key debate within film studies, highlighting questions of film policy, exhibition, programming, reception and audiences' everyday experience of cinematic modernity. This collection presents a range of research methodologies and perspectives from both qualitative and quantitative traditions in order to illuminate the widely varying ways in which cinema has been accepted, rejected or disciplined as an agent of modernity in neighbouring parts of Europe, and to examine how cinema flourished in different European societies: rural, urban, capitalist, communist; as a product and catalyst of modernity, and a provocation of cultural transformations and exchange.

The editors would like to thank all the participants and in particular the keynote speakers, Robert C. Allen and Annette Kuhn, for contributing to the success of the 'Glow in their Eyes' conference, along with Deb Verhoeven, Kate Bowles, Lies Van de Vijver, Gert Willems and Carl de Keyser, the members of the Centre for Cinema and Media Studies at Ghent University, and the Visual Studies and Media Culture Research Group at the University of Antwerp. We are grateful to the Flemish Scientific Research Council (FWO-Vlaanderen), the university film-club Film-Plateau (Ghent

University), and the City of Ghent for their generous support. We also wish to record our gratitude to our contributors for their patience, and to Kathleen Lotze who, together with Khaël Velders, compiled the index.

At Routledge, Natalie Foster and Ruth Moody have steered this project to fruition.

Daniel Bitereyst, Richard Maltby and Philippe Meers

1

CINEMA, AUDIENCES AND MODERNITY

An introduction

Daniel Biltereyst, Richard Maltby and Philippe Meers

> With this system which consists, before going into a movie theatre, of never looking
> to see what's playing – which, moreover, would scarcely do me any good, since
> I cannot remember the names of more than five or six actors – I obviously run the risk
> of missing more than others, though here I must confess my weakness for the most
> absolutely absurd French films.
>
> André Breton[1]

In *Nadja* (1928), one of the landmark modernist novels of the 1920s, French surrealist
writer André Breton included a one-page description of how he used to go to the
movies. In a mixture of provocation, nostalgia and humour, Breton talked about a
'system' of cinemagoing which consisted of never looking at the programme before
entering a cinema. Hoping to be surprised by the 'most absolutely absurd' French
pictures, Breton mostly went to low-grade movie houses in popular areas of Paris,
where people talked loudly, gave their opinions about what was happening on the
screen, and where he and his friends could openly eat and drink while consuming a
picture. His account of opening cans, slicing bread, uncorking bottles 'as if around a
table' in a cinema is interesting not only for its typically surrealist provocative stance
or the Surrealists' fascination for the imaginative world of cinema and its merger with
everyday life. In her insightful analysis of this and other surrealist writings on the
experience of cinema, Jenny Lefcourt forcefully argues that Breton's cinemagoing
system and his description of what happened in Parisian lower-class cinemas were not
unique or exaggerated at all.[2] Bringing to life what he had experienced in the years
following the First World War, the paragraph had a nostalgic tone, along with a
strong documentary value in its description of the unwritten rules of popular cinema
culture, which included talking, eating, drinking and much more, and which seemed
to be fading away as 'spectatorial practices' were 'homogenized' along bourgeois rules
of conduct.[3] Breton's description of the cinema of the 'man in the street' as a quite

savage, uncontrolled site thus incorporated a critical disapproval of how cinema in modern society underwent a gradual process of disciplining through capitalist and bourgeois notions of social order, elitism and cultural distinction.[4]

This book is not about Paris in the 1920s, the Surrealists, or the type of elitist, modernist cinema with which they are often associated. Rather, it concentrates on the experiences and social practice of ordinary moviegoing that Breton looked back on so nostalgically. Although the contributions to this volume deal with various historical European contexts, they all address everyday film culture from a perspective that differs from Breton's principally in its investigative rather than nostalgic tone. The case studies in this book are not primarily concerned with film programmes, stars and authors, or with films as meaningful texts, but rather move towards considering their distribution, exhibition and reception. In their effort to understand exhibition practices, spectatorship and the site of consuming movies, the contributions make a conceptual break with the idea of the movie theatre as a closed space where people are immersed in darkness and where they are submerged in the thrills and shocks of the film spectacle. Trying not to look at audiences as figures constructed by the cinematic apparatus or by particular film texts or genres, the authors insist on moviegoing being conceived as a social act performed by people of flesh and blood, who actively engage with movies and with other people, firmly situated within specific social, cultural, historical and spatial confines. This turn to reception, moreover, goes beyond looking at viewing experiences, cinemagoing practices, or programming and exhibition strategies used to attract or influence those audiences. Like Breton's conception of cinema as a shifting site of ideological struggle and discipline, most of the chapters in this book address wider societal, political and other forces in historicizing film spectatorship, moviegoing and exhibition. Their case studies illustrate how practices and discourses about cinema as a dominant form of leisure in the twentieth century were subjected to or influenced by intermediate forces that included both hard and explicitly disciplining authorities such as state film censorship or military film policy, and the softer forms of power exercised by religious organizations or labour movements.

In its assessment of the wider historical conditions of the cinematic experience, this book joins other efforts by film researchers to cope with the social history of cinema. Since the 1990s, questions of the institutional and economic dimensions of film exhibition, of film programming and exhibition strategies, and of other issues concerning audiences' reception of movies, their memories of cinemagoing and their encounter with the site of cinema, have been high on the research agenda.[5] This broad scholarly examination of film reception has gone hand in hand with an empirical, historical and spatial turn in film studies, and with a criticism of what Robert C. Allen has called the persistent ambivalence within dominant strains of film studies 'toward anything that exists outside the text and beyond the edges of the screen'.[6] Scholars working on the 'flesh and blood human beings who go to cinemas to see films' criticized the fact that the film-centredness of arts- and humanities-based film studies usually implied text-based constructions and theory-led conceptualizations of the spectator in which, as Annette Kuhn argued, films were taken 'as the starting point for exploring the cinema–consumer relationship'.[7] Work by Kuhn and others on bottom-up experiences and memories of cinemagoing not only reconfirmed ideas of audience activity, selectivity and power in a historical context, but also underlined the extent to which cinemagoing was remembered as part of the fabric and routine of social life, thereby questioning the relevance of the movies themselves. Asking whether 'cinematic texts really matter', Allen brought forward the (for some provocative) conclusion that 'generally speaking, for many people in many places for a very long span of film history, the cumulative social experience of habitual or even occasional moviegoing mattered more than any particular film they might have seen'.[8]

Acknowledging that these approaches to film exhibition, audiences and cinemagoing experiences offer an alternative mode of studying key issues in film history, ranging from the relevance of the film canon to wider questions of power in the cinematic circuit of cultural production, Richard Maltby has proposed a distinction between 'film history' and 'cinema history'. While the former primarily deals with an 'aesthetic history of textual relations', the latter is concerned with the 'social history of a cultural institution', or with questions of distribution, exhibition, reception and the 'social experience of cinema'.[9] This 'new cinema history', which according to Maltby has now 'achieved critical mass and methodological maturity', involves the usage of several approaches, coming from history, cultural geography, demography, ethnography and other disciplines within the social sciences.[10] While several overviews of US research on film exhibition and reception have been published in recent years, access to much of the European work in the field has been restricted to those who read the local language.[11] One of the aims of this book is to fill in this gap by making some of the most innovative studies on film exhibition structures, programming strategies and audiences available to an English-speaking audience. A second aim is to provide a basis of evidence for examining the similarities and differences between the experiences of historical audiences in Europe and the US.[12]

This book is, however, more than a showcase of new cinema history work originating from different European countries. The contributions also all address questions to do with the relationship between cinema and the experiences of metropolitan urbanity and modernity, which has been one of the key debates in film history and theory in the last two decades.[13] Since the end of the 1980s, film researchers have argued that cinema was both a consequence and a vital component of modern life.[14] Drawing variously on the writings of Charles Baudelaire on the experience of modernity, Georg Simmel's sociology of space, the metropolis and subjective life, or the work of Walter Benjamin and other early writers on film and mass-mediated modernity, proponents of what David Bordwell has called the 'modernity thesis' emphasized the extent to which early movies reflected the shocks, thrills and fragmented experiences of modern urban life, and speculated that this kind of cinema worked as a catalyst for the experience of modern life by influencing human perception.[15] The modernity thesis proposed that the disruptive economic, social and cultural effects of urbanization and industrialization created a state of constant sensory change, nervous stimulation, feverish stress, speed and bodily peril, and that cinema both reflected this state and was a consequence of it, promoting a particular gaze or form of perception. In a summary of this work, Tom Gunning suggested that movies reflected modernity's disruption of social order by representing 'the experience of urban life with its threats and danger', embracing 'modern technology or new environments', focusing on 'female subjectivity', and emphasizing 'the heightened involvement of a viewer in a visual illusion combined with motion'.[16] Referring to modernity as a culture of shocks and cinema as a reflection of this, Gunning's concept of the cinema of attractions identified a type of filmmaking that solicited spectator attention through visual curiosity, exciting spectacle, surprise, newness, fragmentation and splintered montage.[17]

This collection does not want to discard some of the excellent work within this literature on cinema, metropolis and modernity, nor do we want to reduce the arguments and rhetoric of the modernity thesis *ad absurdum*. Many of these arguments, however, have been heavily criticized by leading scholars in the field such as Bordwell, Allen and Noël Carroll. The thesis' propositions about changes in human perception brought about by the condition of modern life have been strongly challenged, as have claims about the role played in this process by motion pictures.[18] Concentrating on a stylistic history of films, Bordwell argued that many early movies did not reflect a 'culture of splintered experience' and questioned the attempt to tie stylistic aspects of early cinema to modern experience and the fragmentation of urban life.[19] In a recent book on modernity and early cinema, Joe Kember summarized this criticism by arguing that the modernity thesis takes a quite restricted view of modernity, and that the proponents of the thesis tend to exaggerate the reflection of urban modernity in movies and pay less attention to alternative styles. Kember provides examples of early filmmakers demonstrating 'a fascination with bucolic rural life' and 'with exotic locations', suggesting a complex interplay between cinema and a more nuanced conception of modernity.[20] In his work on the attractiveness of early cinema, Kember argues that film institutions tended to market modernity by explicitly promoting cinema's capacity to offer spectacular and disruptive material, but at the same time they invested heavily in reaffirming 'normative descriptions of social identity'. Early film shows, he argues, tried to stitch themselves 'into the fabric of everyday life', and a 'substantial part of the enterprise of such shows involved a revalidation of existing concepts of selfhood and community on behalf of audiences'. Rather than breaking down traditional notions of community, Kember suggests, the film industry 'participated fully in modernity's fascination with, and repackaging of, very traditional social forms'.[21] As a result,

> early cinema needs to be seen as a dynamic, responsive environment which developed multiple relationships – sometimes at the same time – with its varied audiences, and which therefore proliferated experiences of intimacy, empathy, curiosity, reassurance and mastery for individual spectators in place of those it was widely accused of undermining.[22]

A key issue in this critical reconsideration of cinematic modernity and its relation to tradition, community and everyday life, is the role of the audience.[23] Summarizing criticism of the modernity thesis' conceptualization of spectatorship, Kathryn Fuller-Seeley and George Potamianos have argued that it tends 'to transform viewers and their culture, the surrounding theatres and streets, into a vast, anonymous, homogeneous, mass audience'.[24] It postulates a phenomenologically conceived and an apparatus-led spectator who, according to Italian film theorist Francesco Casetti, is 'immersed in the spectacle and the environment', in which the 'cinematographic apparatus … encourages a fusion between subject and object and between subject and environment'. This account of spectatorship endows it with a 'double segregation', where the 'spectator cannot physically touch the screen and what is on it', but 'neither can he or she share intimacy with the other spectators'. Elaborating on Foucault's theories on discipline, social control and the Panoptic gaze, Casetti describes modern cinema as an apparatus that renders 'the subjects – particularly their bodies – "docile."' [25] Modernity theorists have also developed textually inscribed notions of spectatorship. Mary Ann Doane, for instance, has argued that the 'form of the films' encouraged 'spectatorial absorption and submission to an irreversible time' and has advanced some significant propositions about the movie theatre's style of architecture as encouraging a new sense of control, order and safety. Interestingly, Doane confronts this view with some paragraphs on the actual 'viewing situation' in turn-of-the-century working-class movie houses, which was, as she admits, chaotic, boisterous, sociable and interactive.[26]

In recent years, proponents of the modernity thesis have also argued for a more complex definition of modernity and have re-evaluated conceptions of cinematic style, cinema's influence on perception, and its reflection of and relation to modernity at large.[27] But the thesis' master paradigm on film spectatorship has as yet remained unchallenged by a confrontation with more grounded empirical evidence on the reception of cinema and on real audiences' cinemagoing practices. One of this book's aims is to stage some scenes from that confrontation. As well as seeking to avoid sweeping generalizations about film audiences by examining them empirically in specific historical terms, this book embraces the idea of a multiple and more dynamic definition of modernity. This definition refers to critical social theory's account of the ambivalences of modernity, in which counter-forces or alternative traditions of modernity compete with hegemonic or culturally dominant forms of it.[28] While this more dynamic model is one of the recurrent items in recent re-assessments of the cinema/modernity thesis, it remains unclear how this might be operationalized in empirical historical film research.[29] The chapters of this book present a range of positions on the cinema–modernity thesis, and it is important to begin by acknowledging that some ideas and insights tend to confirm aspects of the thesis' view. Annemone Ligensa's chapter on Berlin elaborates on the metropolis as the site of cinema and nervous relentlessness, while Andrea Haller's account of '*Flimmeritis*' in the German pre-First World War period emphasizes the role of film theatres as a site of femininity, and explores contemporary views of cinema as a catalyst of modern lifestyles and values and a possible threat to the social order. Such arguments were as commonly made in Europe as they were in the USA during the public debates, anxieties and moral panics over cinema's emergence and popular appeal. Several essays highlight key arguments in these *Kino-Debatten*: Haller examines anxieties over the visibility and the gendered transformation of the public sphere, and Daniel Biltereyst, Philippe Meers, Kathleen Lotze and Lies Van de Vijver chronicle the Roman Catholic church's well-organized attacks on the devastating effects of 'sex-appeal' in American movies in Belgium. The latter refers to another, typically European recurrent item in cinema–modernity debates, namely that over the dangers of American cinema and the disrupting values it supposedly carried with it, such as consumerism, violence and horror.[30] Throughout the book contributors illustrate how cinema's modernity has been debated in every centre of Europe in ways which reflect local imprints on modernism and modernity.[31]

The collection, however, offers many arguments that run against the grain of the cinema–modernity–metropolis thesis. One level of this antithesis moves in parallel with research from the USA that has decentralized the metropolitan focus on the experience of (early) cinema.[32] The specific rural, small town and regional town geographies of film culture are examined in Anna Manchin's account of exhibition in Hungary, Thunnis Van Oort's description of exhibitors as cultural brokers in the southern part of the Netherlands, and Biltereyst, Meers, Lotze and Van de Vijver's examination of rural cinemas in Belgium, while Åsa Jernudd argues that in Sweden, cinema developed more intensively in small towns than in a metropolitan environment. Another level at which these essays question the modernity thesis in many ways takes place in their examination of audiences' experiences and cinemagoing practices. Very much in line with debates on the relativist stance of (qualitative) audience research in other domains of media studies, we could interpret these accounts of actual audiences as highlighting competent, active and selective consumers, who playfully 'embraced the various attractions on offer without fainting or becoming hysterical', as Haller writes in her description of female film fans as self-confident film connoisseurs, interpreting films according to their own preferences, 'laughing when others were crying'.[33] Other chapters, including Pavel Skopal's examination of post-war communist Czechoslovakia, contain examples of resistant or negotiating viewers not being immersed by the thrills, shocks or drifts in movies and cinemas, up to the point of audiences staying away from cinemas.

On a macro level, a strong argument against the modernity thesis consists in considering the sweeping hegemonic resistance to cinema's modernity that emerged all over Europe. Probably more than any other art form or medium in the twentieth century, cinema has been the object of harsh control, either in the straightforward form of censorship or other, more complex forms of discipline. The chapters in this book illustrate how in very different historical contexts, various societal, political or ideological forces tried to discipline not only the production but also the free distribution and exhibition of movies, along with attempts to 'guide' audiences in their cinemagoing habits. Although these interventions were often inspired by local imperatives, many of the intervening forces responded to anxieties occasioned by cinema's modernity, deploying discourses preoccupied with the protection of children, values and norms, and with the restoration of traditions and community life as well as nationalism and national belonging. Among those disciplining forces, the book highlights government interventions and state film policy (Skopal and Petr Szczepanik on attempts to limit

American and other Western movies in communist Czechoslovakia; Manchin on Hungary), film censorship systems (Jernudd; Manchin; Biltereyst, Meers, Lotze and Van de Vijver), military forces (Frank Kessler and Sabine Lenk on cinemagoing in occupied Düsseldorf), initiatives by religious groups or institutions (Jernudd; Van Oort; Biltereyst, Meers, Lotze and Van de Vijver), grassroots, workers' and other progressive movements (Jernudd, Moitra), as well as the film industry itself (Haller, Van Oort). This view of cinema as a site of ideological struggle, where even in a free market system cinematic modernity is confronted by forces representing counter- or even anti-modern values, very much corresponds with wider historical-political analyses of twentieth-century Europe as a laboratory of political and ideological confrontation. Developing ideas of 'multiple modernities', the chapters in the book exemplify various types of counter- or alternative forms ranging from notions of provincial modernity (Jernudd), Fascist-inspired modernity (Manchin) to socialist modernity (Skopal, Szczepanik).[34]

It is in this context that the most obvious difference framing European and American experiences of cinema – that of nationality – can most usefully be considered. National cinema has most commonly been understood as a defining category of film production, most commonly seen as not merely distinguishing smaller domestic production industries from Hollywood's international reach, but placing them firmly in opposition to a globalizing American culture. Cinema history is still absorbing the resonances of Andrew Higson's argument that a discussion of national cinema must necessarily consider the films that national audiences consumed.[35] In this respect, contributions to this book provide evidence that indicates audiences' long-term preferences for the culturally familiar, along with an ambivalent response to imported cultural products. Peter Szczepanik's account of Czech audiences' tastes, for example, chimes well with Lawrence Napper's argument that the aspiration to use cinema to constitute an integrated national culture in 1930s Britain was most successfully fulfilled by consensual, 'middlebrow' forms of entertainment.[36] More generally, the work in this book argues that specific, localized case studies demonstrate the diversity of audience experiences, and that any assessment of cinema's relationship with modernity must acknowledge the range of that diversity. From this perspective, national differences may be seen as only one element in the determinants of audience experience, and other determinants may draw attention to other patterns of difference: differences between regions within or across national boundaries, for example. By the same token, the evidence presented in these contributions suggests that it would be fruitful to consider the extent to which small-town communities in Europe and the US shared similar experiences of cinema, distinct in character and social function from those of urban audiences.

In keeping with these suggestions, this book tries to move beyond a confrontational model and embraces ideas of modernity as a site of exchange and negotiation. Again, different levels of evidence and argument can be identified. Several chapters, including Jernudd's, examine the relationship of exchange and mutual influence between the metropolis, small towns and rural environments, highlighting the idea that 'the modern' is not solely to be interchanged with 'the urban'.[37] Kessler and Lenk's chapter illustrates the complexity of the forces disciplining cinema's modernity: even in the case of a military film policy in German occupied areas, 'political authorities, exhibitors, and audiences had to negotiate the different forces and interests that acted upon the experience of movie-going'. A key role in this negotiation was played by exhibitors acting as intermediaries or cultural brokers. Van Oort's chapter details the important role of these 'individual actors (both persons and organizations) in integrating cinema culture into a social and cultural context, framing the way that entrepreneurs and policymakers negotiated the specific characteristics of cinema cultures'. Descriptions of the processes of cultural exchange, negotiation and mixing also appear in Pierluigi Ercole's account of the role played by cinema in the identity of diasporic populations, such as the Italian community in London in the First World War. These processes are also central to several contributors' examinations of audience behaviour. Most of the chapters touch on questions of the cinemagoer's autonomy and power in their interchange with cinema's modernity. In their discussion of how audiences experienced the interventions of disciplining forces such as the censors and the Roman Catholic church, Biltereyst, Meers, Lotze and Van de Vijver argue that these had often an inverse effect, so that people tended to be disproportionately attracted by forbidden movies. In her chapter on female cinemagoers, Haller indicates how women were both attracted by stars and tried to imitate them, but equally transformed the mythologies of stardom for purposes of self-definition and the formation of a (temporary) self-identity.

The first part of the book focuses on the wider historical conditions of the cinematic experience and the disciplining forces trying to control it in the name of traditional and community values. Beginning with an analysis of the literature describing the modernization of Swedish society, Åsa Jernudd's chapter concentrates on early cinemagoing in rural Sweden, and specifically on film exhibition, programming, advertising and cinemagoing practices in the small town of Örebro. Jernudd argues that in its first decade, film exhibition in Sweden was primarily a feature of small-town, organized lower middle- and working-class life, rather than a metropolitan experience. In stark contrast to the situation in the cities of Stockholm and Gothenburg, where film exhibition declined after its initial novelty appeal, the transformation of itinerant film exhibition to permanent forms was a gradual and relatively inconspicuous process in rural and small-town Sweden. Underlining the ease with which film exhibition was received, Jernudd looks at how diverse popular and oppositional social movements, grassroots and labour organizations incorporated film exhibition into already existing forms of entertainment, and how they were the key to understanding cinema's social acceptance and growing cultural legitimacy. Illustrating provincial modernity's attempts to accommodate tradition, her chapter shows how cinema was part of a broader movement of social progress and mediation between the old and the new, literally providing a space of transformation.

In their chapter on moviegoing under military occupation, Frank Kessler and Sabine Lenk focus upon the interplay between cinema, audiences, and political and military power. At the end of the First World War, the regions on the western banks of the Rhine were occupied by Belgian, British and French forces. The occupied zones included the part of Düsseldorf situated on the western bank of the river, so that the town was partly occupied by Allied forces. This resulted in a complex administrative situation, with different authorities being involved in regulating public life, including cinema. Using a wealth of archival material, Kessler and Lenk explore the impact of this situation on the functioning of local cinemas and more generally on moviegoing practices. They examine the effect that the division of the town had on the moviegoing public, the varying regulations and regimes of censorship, and the different programmes shown in the cinemas on both sides of the Rhine. They consider the role of the occupying administration in controlling programmes, providing special shows for the occupying forces, and privileging imports from the occupying countries. This specific case study provides a unique illustration of the complex interplay between political and other forces shaping a given historical situation, and the ways in which the entertainment sector adjusted to it as it aimed to offer their audiences the pleasure they sought.

Concentrating on the southern part of the Netherlands in the 1910s and the 1920s, Thunnis Van Oort addresses the important role played by Roman Catholic organizations in film exhibition, film culture and cinemagoing. The Catholic perspective provides a significant insight into the Dutch experience of cinema, because Catholic communities were organized, in varying degrees, at local, regional, national and international levels. The chapter elaborates on the Catholic influence on the film industry, the establishment of a regional Catholic censorship organization, and the crucial role played by cinema owners or exhibitors who acted as cultural brokers, deciding how audiences were exposed to modern international film fare. Using the concept of acculturation, Van Oort argues that film exhibitors had to negotiate between the demands of the audience, the interests of the national and international distributors supplying films, and a range of ideological and regulatory pressures from government and Catholic organizations.

Anna Manchin's chapter deals with film culture in Hungary during the interwar years, when films were extremely popular with local audiences in both urban areas and the countryside. In Hungary, as in most other parts of continental Europe, films were widely available and accessible to almost everyone, and by the interwar years, they had become arguably the most important form of mass culture. The new public sphere represented by cinema was also accorded an immensely important propaganda role in defining Hungarian national culture and the new Hungarian national community. Conservative government circles and radical nationalist critics alike regarded the modernizing potential of films as a grave threat to traditional Hungarian culture and to rural culture in particular. In the late 1930s, the increasingly Fascist Hungarian government tried to prevent this destructive effect by limiting the number of theatre licences in small rural areas, by attempting to create films showcasing 'true' Hungarian national culture, and by excluding Jewish filmmakers from the Hungarian film industry. Liberal intellectuals, on the other hand, welcomed the effects of mass culture on Hungarian society, believing that the spreading of urban popular culture to the countryside was an important way of extending modern, liberal, democratic values to society at large. The chapter explores the consequences and cultural confrontations that these contesting views of the place of cinema provoked in interwar Hungarian culture and society.

In his chapter on post-war communist Czechoslovakia, Pavel Skopal focuses on the 40-year-long era when the state made direct use of cinema as a vehicle for political and ideological goals. Arguing that cinema played a significant role in the socialist project of modernization, Skopal analyses the cultural politics of a communist system that used film distribution and exhibition as an instrument of education and in which 'socialist man' was defined as a new kind of film viewer. These mechanisms of distribution and practices of exhibition created an historically quite specific set of relationships between the state, the film industry, and the citizen-viewer. Using materials from various Czech archives, Skopal argues that post-war communist Czechoslovakia exemplifies the 'schizophrenic' mechanism of planned distribution and attendance, which found itself incapable of achieving either economic or ideological efficiency, or of fully satisfying audiences' cultural preferences.

In his chapter on South Wales miners' cinemas in the 1950s and 1960s, Stefan Moitra looks at a very particular kind of cinema operation in an industrial region. Initially following a programme of promoting political consciousness and 'high culture' to their working-class clientele, the Welsh miners' institutes started to use cinema from at least the 1920s. After a brief overview of their pre-war history, the chapter concentrates on the post-war period, and the continuing problem created by the tension between the institutes' educational agenda and the financially more advantageous provision of commercial entertainment. Highlighting the place of cinema in these mining communities and in the process of their post-war modernization, Moitra also discusses the role of the miners' institutes as entrepreneurs in the cinema industry, and the crisis and subsequent closure of working-class film theatres in the 1960s.

While the chapters in Part I mainly discuss structural attempts to control the historical conditions of the cinematic experience under the flagship of national, traditional and community values, the contributors to Part II deal with various processes by which audiences actively engaged with the world of cinema and negotiated their identities through the activity of cinemagoing. Concentrating on the pre-war period, usually considered to be the mainland of high modernity, most of the chapters focus on cultural exchange, both on an individual and a collective level, as well as on the interchange of persons, movies and other commodities and the values they carry with them. This more dynamic model of interaction with cinematic novelties also considers issues of distribution, the importation of foreign movies, the role of audiences in engaging with new forms of cinema, and the role of cinema in diasporas. Arguing that the modern world of cinema need not be regarded as an alienating feature of modern life, these chapters all illustrate the ability of audiences and communities to accommodate and integrate various aspects of modernity. Annemone Ligensa's opening chapter on cinemagoing in pre-First World War Germany argues that the relationship between early cinema, urbanization and modernity is much more complex and indirect than contemporary critics and current modernity theory frequently imply. From Weber, Simmel and Benjamin onwards, cultural criticism has conventionally linked cinema with urbanization, projecting an image of the film spectator as an anonymous and isolated figure, fixated on the screen. Ligensa reconsiders this conception of the prototypical cinemagoer as a neurasthenic city dweller. Using Germany between 1895 and 1914 as her case study, she argues that in the light of empirical knowledge about actual audience activity, this conception is more dystopian myth than useful description. On the level of individual behaviour, sociologist Emilie Altenloh's study of the German industrial city of Mannheim in 1912 already showed that there were no significant differences in cinema attendance between native or long-time inhabitants and recent migrants or commuters from the country. Using the databases of the Siegen research project and beginning in 1905, Ligensa describes the development of the German cinema scene, indicating how permanent cinemas appeared and spread from large cities, as well as how they increasingly displaced travelling cinemas in small towns.

Andrea Haller's chapter also starts from Altenloh's landmark study of film audiences, and goes on to reconstruct the discourses on female cinemagoing in Imperial Germany. Using contemporary trade papers, fan and women's magazines, Haller examines how women experienced their moviegoing and participated in the actual event of the film show, and how the patriarchal society to which they belonged reacted to their participation in this new activity. This examination of the formation of a female film-fan culture and the female moviegoing experience shows that going to the cinema had become an important part of women's lives during the first two decades of the twentieth century, of more consequence than simply providing them with access to entertainment. Experienced as a means of partaking in modern life, cinema offered women the chance to 'see themselves' and to acquaint themselves with different female lifestyles. The chapter argues that the public commentary on female cinemagoing can be read as a discourse on visibility: on how women became visible in public and in the auditorium, on what was not visible in the dark and on what was suspected to happen there, as well as on what women could have watched on screen.

One of the movies they might have watched was the controversial *Afgrunden* (*The Abyss*, 1910), one of the first feature films in Germany and responsible for launching the career of Danish film star Asta Nielsen. In his case study of the movie's circulation Martin Loiperdinger explicitly uses an audience perspective to demonstrate how the shift from the short film programme to the long feature film was fundamental to the establishment of the commercial film programme format. The role of audiences has so far been relatively little examined in the literature on the emergence of the feature film. The feature film was introduced to the German film market in late 1910 through Ludwig Gottschalk's creation of the Monopolfilm distribution strategy in his marketing of *Afgrunden*, which was a great success in Germany. Cinema historiography has tended to describe this breakthrough by using sources in the trade press that concentrate on film distribution and advertising. In contrast, Loiperdinger argues that the enthusiasm of cinemagoers for the long feature was crucial to *Afgrunden*'s success. Exploring previously unexamined sources from various local newspapers and archives, the chapter reconstructs the patterns of audience attendance and audience response to the film, detailing its crucial role in the decision, taken in spring 1911 by German film businessmen, to promote Nielsen as the first European film star.

In the following chapter, Pierluigi Ercole introduces issues of cinema, diaspora, nationalism and national belonging into his consideration of the involvement of Italian immigrants in the screening of Italian war films in London during the First World War. Starting in 1915, two Italian immigrants involved in the British film industry, Mario Pettinati and Arrigo Bocchi, began to organize special events targeting the Italian community in London. As well as reinforcing a sense of national identity and belonging among the immigrant colony, these screenings showed the British public the war effort of their Italian allies. Mirroring the nationalistic discourses that pervaded the London-based Italian press of the time, Pettinati and Bocchi's screenings aimed to promote national values derived from the history of Italy's political and geographical unification since 1861. Looking specifically at the cultural activities organized by members of Italian clubs and societies, Ercole investigates how the concept of national belonging was questioned by the working-class section of the colony, and discusses the reception of Italian war films and their propagandist message within the diasporic social context of London's Italian colony.

Petr Szczepanik's chapter concentrates on another form of cinematic cultural exchange, in the importation, exhibition and reception of foreign movies in the Czechoslovak film market in the 1930s. Using a sample of over 2,500 films premiered in 20 first-run movie theatres in Prague (including multiple-language and dubbed versions from German, American and other sources), Szczepanik analyses the short-term popularity of American talkies in the period immediately after the introduction of sound, and the marked decrease in their popularity in the years that followed. He examines why both the English language and American culture came to be regarded as disruptive elements by local audiences, and explores which kinds of American films continued to be successful after 1930. He also discusses whether it was the German language, which was much more comprehensible to the local public than English, that made German films more popular than American ones, or whether it was their expression of the archetypes of German-Austrian popular culture that made them more successful than not only Hollywood's products, but also the German-speaking versions of American films. Invoking Fernand Braudel's 'dialectic of duration' through which he sought to describe the simultaneous plurality of historical time, Szczepanik also points to a method by which we can use Braudel's distinction between the cyclical, social time of 'the major forms of collective life' and the individual time of particular events to establish a temporal framework within which we can more exactly examine patterns of cultural consumption.[38]

The final chapter of this collection is based on a large-scale research project on the history of film exhibition, programming and audiences' cinemagoing experiences in the Northern part of Belgium, the Dutch-speaking region of Flanders. In this chapter Daniel Biltereyst, Philippe Meers, Kathleen Lotze and Lies Van de Vijver bring together questions of control and resistance to cinema's modernity with issues of audiences' experiences with it. In the first part of the chapter, the authors indicate how various ideological–political groups developed strategies by which to influence film exhibition and cinemagoing practices, mainly by exploiting a network of associated film theatres. As in the Netherlands, a powerful block of Catholic organizations came to control a large chain of cinemas and other viewing spaces, mainly in rural areas; at the same time, in a manner comparable to the Legion of Decency in the US, they also installed a system of film classification for those film theatres. From 1921 onwards, Belgium also faced a non-obligatory system of state film censorship, which proved on a daily basis to be severe in cutting images and limiting the screening of movies that were considered to be harmful to children under 16. The chapter goes on to examine the efficiency of these disciplining forces. Questioning former regular film viewers on how they experienced censorship, the authors come to the conclusion that people were very much aware of these forces and frequently tried to reject or escape them, up to the point that systems of control could have an inverse effect by promoting forbidden or censored movies.

Collectively, the aim of the essays in this book is not only to bring together or highlight recent European case studies in the field of new cinema history. In many ways, the main concern of this collection is to emphasize how sensitive empirical studies on everyday media distribution, exhibition, consumption and reception can contribute to film history and grand theories within the field, as they have done in other domains of media culture research.[39] Acknowledging that the analysis of the interconnections between cinema, modernity and metropolitan urbanity is still powerful in explaining particular shifts in modern cinematic style and experience, this book argues that empirical evidence on the reception side of the spectrum complements, enriches and to some degree complicates our understanding of these matters. As part of what Ben Singer has recently called the current critical–historical framework for thinking about cinema and modernity, the contributions to this book argue that cinema's modernity goes much further than style and other textual related issues, or questions on perception and spectatorship.[40] Underlining the need to take into account wider historical conditions of the cinematic experience, this book aims to contribute to a complex social history of cinema. The chapters emphasize the extent to which individual viewers, wider communities or other ensembles of people did not simply and submissively undergo cinema. Besides exchange and negotiation, both on an individual and a collective level, society reacted and discussed modernity in its multiple appearances. Eventually, public debates and moral panics led to harsher forms of control, resistance or even censorship. This perspective produces a picture of cinema as a complicated arena in which modern, anti-modern and alternative forms of modernity met. Cinema became a site of struggle reflecting wider processes of ideological competition, and in this regard, twentieth-century Europe has been more than just a unique experiment.

Notes

1 A. Breton, *Nadja*, New York: Grove Press, 1960 (orig. 1928, trans. Richard Howard), p. 37.
2 J. Lefcourt, 'Aller au cinéma, aller au peuple', *Revue d'Histoire Moderne et Contemporaine* 4, 2004, 98–114.
3 M. Richardson, *Surrealism and Cinema*, Oxford: Berg, 2006. See particularly chapter 5, 'Surrealism and Documentary'.

4 Quoted from French poet Philippe Soupault, in Lefcourt, 'Aller au cinéma, aller au peuple', p. 105. This 'embourgeoisement' not only referred to attempts to attract higher social classes to the cinema, but also encompassed notions of elitism and social distinctions, a cinéphile defence of cinema as an art form, the promotion of particular movies and efforts to improve the movies' morality through censorship.

5 R.C. Allen, 'From exhibition to reception: reflections on the audience in film', *Screen* 31, 1990, 347–56; B. Klinger, 'Film History Terminable and Interminable: Recovering the Past in Reception Studies', *Screen* 38, 1997, 107–28; R.C. Allen, 'Relocating American Film History: The "Problem" of the Empirical', *Cultural Studies* 20, 2006, 48–88; G.A. Waller, *Main Street Amusements: Movies and Commercial Entertainment in a Southern City, 1896–1930*, Washington: Smithsonian Institution Press, 1995; G.A. Waller (ed.), *Moviegoing in America: A Sourcebook in the History of Film Exhibition*, Malden, MA: Blackwell, 2002; K.H. Fuller, *At the Picture Show: Small-town Audiences and the Creation of Movie Fan Culture*, Charlottesville: University Press of Virginia, 1996; M. Stokes and R. Maltby (eds), *American Movie Audiences: From the Turn of the Century to the Early Sound Era*, London: British Film Institute, 1999; M. Stokes and R. Maltby (eds), *Identifying Hollywood's Audiences: Cultural Identity and the Movies*, London: British Film Institute, 1999; M. Stokes and R. Maltby (eds), *Hollywood Spectatorship: Changing Perceptions of Cinema Audiences*, London: British Film Institute, 2001; J. Sedgwick, *Popular Filmgoing in 1930s Britain: A Choice of Pleasures*, Exeter: University of Exeter Press, 2000; M. Jancovich, L. Faire and S. Stubbings, *The Place of the Audience*, London: BFI, 2003; R. Maltby and M. Stokes (eds), *Hollywood Abroad: Audiences and Cultural Exchange*, London: BFI Publishing, 2004; J. Najuma Stewart, *Migrating to the Movies: Cinema and Black Urban Modernity*, Berkeley: University of California Press, 2005; R. Abel, *Americanizing the Movies and 'Movie-mad' Audiences, 1910–1914*, Berkeley: University of California Press, 2006; R. Maltby, M. Stokes and R.C. Allen (eds), *Going to the Movies: Hollywood and the Social Experience of Cinema*, Exeter: University of Exeter Press, 2007; K.H. Fuller-Seeley (ed.), *Hollywood in the Neighborhood: Historical Case Studies of Local Moviegoing*, Berkeley: University of California Press, 2008.

On reception studies and historical spectatorship, see J. Staiger, *Interpreting Films: Studies in the Historical Reception of American Cinema*, Princeton, NJ: Princeton University Press, 1992. See also M. Hansen, *Babel and Babylon: Spectatorship in American Silent Film*, Cambridge, MA/London: Harvard University Press, 1991; J. Staiger, *Perverse Spectators: The Practice of Film Reception*, New York: New York University Press, 2000.

On cinema historiography and popular memory, see J. Mayne, *Cinema and Spectatorship*, London: Routledge, 1993; J. Stacey, *Star Gazing: Hollywood Cinema and Female Spectatorship*, London: Routledge, 1994; A. Kuhn, *An Everyday Magic: Cinema and Cultural Memory*, London: I.B. Tauris, 2002 (the US edition: *Dreaming of Fred and Ginger: Cinema and Cultural Memory*, New York: New York University Press, 2002); V. Burgin, *The Remembered Film*, London: Reaktion Books, 2004.

6 R.C. Allen, 'The Place of Space in Film Historiography', *Tijdschrift voor Mediageschiedenis* 9, 2006, 15. On this empirical-historical turn, see R.C. Allen and D. Gomery, *Film History: Theory and Practice*, New York: McGraw-Hill, 1985; 'In Focus: Film History, or a Baedeker Guide to the Historical Turn', *Cinema Journal* 44, 2004, 94–143.

7 A. Kuhn, *An Everyday Magic*, pp. 3–4.

8 R.C. Allen, 'Relocating American Film History', 59–60.

9 R. Maltby, 'How Can Cinema History Matter More?', *Screening the Past* 22, December 2007, www.latrobe.edu.au/screeningthepast/22/board-richard-maltby.html; R. Maltby, 'On the Prospect of Writing Cinema History from Below', *Tijdschrift voor Mediageschiedenis* 9, 2006, 85.

10 See Maltby's 'Introduction' to *Explorations in New Cinema History*, R. Maltby, D. Biltereyst and Ph. Meers (eds.) (Malden, MA: Blackwell, 2011). See also K. Bowles, R. Maltby, D. Verhoeven and M. Walsh, *The New Cinema History: A Guide*, Malden, MA: Blackwell, forthcoming.

11 See S.J. Ross (ed.), *Movies and American Society*, Malden, MA: Blackwell, 2002; Waller, *Moviegoing in America*, 2002; Fuller-Seeley, *Hollywood in the Neighborhood*, 2008.

12 Some large-scale research projects make available (parts of their) data on the internet. See the Czeck *Film Culture in Brno (1945–1970)* project (www.phil.muni.cz/dedur/index. php?&lang=1), the Dutch *Cinema in Context* database (www.cinemacontext.nl/) or the *London Project* (londonfilm.bbk.ac.uk/).

13 Following Thomas Elsaesser, we distinguish between 'modernism' in the 'sense of an artistic avant-garde', different forms of 'modernization' as they affect labour and work 'in technology, industry and science', and 'modernity' as a practice of 'lifestyle, fashion and sexual mores', and as referring to a 'particular attitude to life, in Western societies usually associated with increased leisure time and new patterns of consumption' (Th. Elsaesser, *Weimar Cinema and After: Germany's Historical Imaginary*, London: Routledge, 2000, 390).

14 This position is best summarized in chapter 4 of Ben Singer's *Melodrama and Modernity: Early Sensational Cinema and its Contexts* (New York: Columbia University Press, 2001). See also L. Charney, *Empty Moments. Cinema, Modernity, and Drift*, Durham, NC: Duke University Press, 1998; L. Charney and V.R. Schwartz (eds), *Cinema and the Invention of Modern Life*, Berkeley, Los Angeles, London: University of California Press, 1995; M.A. Doane, *The Emergence of Cinematic Time. Modernity, Contingency, the Archive*, Cambridge, MA: Harvard University Press, 2002; C. Keil and S. Stamp (eds), *American Cinema's Transitional Era*, Berkeley: University of California Press, 2004; F. Cassetti, *Film, Experience, Modernity*, New York: Columbia University Press, 2005; M. Pomerance (ed.), *Cinema and Modernity*, New Brunswick, NJ: Rutgers University Press, 2006.

15 D. Bordwell, *On the History of Film Style*, Cambridge, MA: Harvard University Press, 1997, 141–46. Seminal texts are Charles Baudelaire (1863), *The Painter of Modern Life*; George Simmel (1903), *The Metropolis and Mental Life*; Walter Benjamin (1936), *The Work of Art in the Age of Mechanical Reproduction*, all reproduced in V.R. Schwartz and J.M. Przyblyski (eds), *The Nineteenth-Century Visual Culture Reader*, New York: Routledge, 2004.

16 T. Gunning, 'Early American film', in J. Hill and P. Church Gibson (eds), *The Oxford Guide to Film Studies*, Oxford: Oxford University Press, 1998, p. 266.

17 See on this, T. Gunning, 'The Cinema of Attractions: Early film, its Spectator and the Avant-garde', *Wide Angle* 8, 1986, 63–70; T. Gunning, 'Early American film', 1998; T. Gunning, 'Modernity and Cinema: A Culture of Shocks and Flows', in M. Pomerance (ed.), *Cinema and Modernity*, 2006, pp. 297–315.

18 See on this, N. Carroll, 'Modernity and the Plasticity of Perception', *The Journal of Aesthetics and Art Criticism* 59, 2001, 11–17. Bordwell, *On the History of Film Style,* pp. 144–46.

19 Bordwell, *On the History of Film Style,* pp. 143, 145.

20 J. Kember, *Marketing Modernity: Victorian Popular Shows and Early Cinema*, Exeter: University of Exeter Press, 2009, p. 18.

21 Ibid., pp. 213–14.

22 Ibid., p. 213.

23 See also R.G. Walters, 'Conclusion: When Theory Hits the Road', in K.H. Fuller-Seeley, *Hollywood in the Neighborhood,* pp. 253–54.

24 K.H. Fuller-Seeley and G. Potamianos, 'Introduction: Researching and Writing the History of Local Moviegoing', in *Hollywood in the Neighborhood*, p. 5.

25 Casetti, *Film, Experience, Modernity*, pp. 166, 176.

26 Doane, *The Emergence of Cinematic Time,* pp. 132–33.

27 See contributions by Gunning, Keil, Singer and others in *American Cinema's Transitional Era,* 2004; Gunning, 'Modernity and Cinema: A Culture of Shocks and Flows', 2006; contributions by Gunning, Musser and Keil in W. Strauven (ed.), *The Cinema of Attractions Reloaded*, Amsterdam: Amsterdam University Press, 2006; contributions by Singer, Gunning and others in *Film 1900: Technology, Perception, Culture*, 2009. See also Kember, *Marketing Modernity*.

28 F. Jameson, *Postmodernism, or The Cultural Logic of Capitalism*, Durham, NC: Duke University Press, 1991.

29 In a recent debate, for instance, Ben Singer launched the concept of the 'ambimodernity' of early cinema. See B. Singer, 'The Ambimodernity of Early Cinema: Problems and Paradoxes in the Film-and-Modernity Discourse', in A. Ligensa and K. Kreimeier (eds), *Film 1900: Technology, Perception, Culture,* New Barnet, Herts.: John Libbey, 2009, pp. 37–39. For an earlier discussion, M. Bratu Hansen, 'America, Paris, the Alps: Kracauer (and Benjamin) on, Cinema and Modernity', in *Cinema and the Invention of Modern Life,* 1995, pp. 362–402.

30 For an audience-oriented perspective on foreign (including European) engagements with American cinema, see M. Stokes and R. Maltby (eds), *Hollywood Abroad: Audiences and Cultural Exchange,* London: British Film Institute, 2004. See also R. Maltby and R. Vasey, '"Temporary American Citizens": Cultural Anxieties and Industrial Strategies in the Americanisation of European Cinema', in *'Film Europe' and 'Film America': Cinema, Commerce and Cultural Exchange, 1925–1939,* Exeter: University of Exeter Press, 1999, pp. 32–55; V. de Grazia, *Irresistible Empire: America's Advance through Twentieth-Century Europe,* Cambridge, MA: Harvard University Press, 2005.

31 On German cinema debates, see: A. Kaes (ed.), *Kino-Debatte: Texte zum Verhältnis von Literatur und Film 1909–1929,* Tübingen: Max Niemeyer Verlag, 1978. On the Spanish anti-cinema movement in the 1910s and 1920s, where both intellectuals and the church attacked cinema, see J. Minguet Batllori, 'Early Spanish cinema and the problem of modernity', *Film History* 16, 2004, 95–98. For Czech theories on cinema and modernity, see J. Andel and P. Szczepanik (eds), *Cinema All the Time: An Anthology of Czech Film Theory,* Prague: National Film Archive, 2008. On modernity, modernism and film exhibition in France, see C. Gauthier, *La Passion du Cinéma,* Paris: AFRHC, 1999.

32 G.A. Waller, *Main Street Amusements,* 1995; K.H. Fuller, *At the Picture Show,* 1996; R.C. Allen, 'Relocating American Film History', 2006, 62–64; K.H. Fuller-Seeley, *Hollywood in the Neighborhood,* 2008.

33 See the debate on (qualitative) audience research in media and cultural studies, starting with M. Morris, 'Banality in Cultural Studies', *Block* 14, 1988, 15–25; D. Morley, *Television, Audiences, and Cultural Studies,* London: Routledge, 1992; chapter 2 in I. Ang, *Living Room Wars: Rethinking Media Audiences for a Postmodern World,* London: Routledge, 1996; M. Ferguson and P. Golding, *Cultural Studies in Question,* London: Sage, 1997. For a wider discussion on audiences and critical media studies, see D. Biltereyst and Ph. Meers, 'The political economy of audiences', in J. Wasko, G. Murdock and H. Sousa (eds), *The Handbook of Political Economy of Communications,* Malden, MA: Blackwell, 2011, pp. 415–435.

34 See J. Jenkins, *Provincial Modernity,* Ithaca, NY: Cornell University Press, 2003. See also Fuller-Seeley, *Hollywood in the Neighborhood,* 2008; E.C. Carlston, *Thinking Fascism: Sapphic Modernism and Fascist Modernity,* Stanford, CA: Stanford University Press, 1998; S. Reid and D. Crowley (eds), *Style and Socialism: Modernity and Material Culture in Post-war Eastern Europe,* Oxford: Berg, 2000.

35 A. Higson, 'The Concept of National Cinema', *Screen* 30:4 (Autumn 1999), 36–46.

36 L. Napper, *British Cinema and Middlebrow Culture in the Interwar Years,* Exeter: University of Exeter Press, 2009.

37 For a Marxist-inspired analysis of this, see N. Zweig, 'Foregrounding Public Cinema and Rural Audiences: The USDA Motion Picture Service as Cinematic Modernity, 1908–38', *Journal of Popular Film and Television* 37, 2009, 116–25.

38 F. Braudel, *On History,* Chicago: University of Chicago Press, 1980, pp. 10–11, 26.

39 See K.H. Fuller-Seeley and G. Potamianos, 'Introduction: Researching and Writing the History of Local Moviegoing', in *Hollywood in the Neighborhood,* 2008, 3.

40 Singer, 'The Ambimodernity of Early Cinema', 37.

Part I
Cinema, Tradition and Community

2

SPACES OF EARLY FILM EXHIBITION IN SWEDEN

1897–1911

Åsa Jernudd

Introduction

Theories of modernization and cinema often suggest that early film culture should be understood in the context of the metropolis, where the *flâneur* and the less respectable *flâneuse* serve as models for spectatorship. This literature, which has until recently remained largely unchallenged in film historiography, claims that disruptive economic, social and cultural forces created a state of constant sensory violence, of nervous stimulation, stress and bodily peril.[1] This may seem viable for film exhibition in its initial decade, which coincided with the exceptionally rapid urbanization and booming industrial economy that characterized Sweden in the 1890s and the early years of the twentieth century. It is, however, reasonable to question the presumed impact of modernism as a 'cultural dominant' or what Fredric Jameson called a wide-spread 'force field'.[2]

This chapter is concerned with early film exhibition in Sweden during the time when itinerant forms of exhibition were dominant and cinema was in the making as a new form of institutionalized mass medium. In keeping with the kind of social historiography that pays attention to spaces of exhibition, I will argue that the introduction of film and the opening of permanent-site exhibition was a gradual process closely intertwined with locally defined social practices, indicating a smooth transition from travelling to permanent forms of exhibition and representing a kind of space that neutralized the anxieties that were elsewhere aroused by the medium's novel and urban appeal. A clarification of the term 'space' is in order. It is adopted from geographer Doreen Massey's theory of space as a social sphere. The idea of social interaction and the concept of time are central to her theory: space is a sphere of interrelations of embedded material practices that are always in the process of being made.[3]

My account differs from the received national film historiography that lends priority to the big cities of Stockholm and Gothenburg, presenting a different conception of the spaces of film exhibition.[4] This chapter will reveal how these two spaces of the early period of film history are constructed differently and in unequal terms. The spaces of film exhibition in the Big City have been used in received film historiography to represent the nation, marginalizing the town as well as the smaller community experience. I share the proposition held by a growing number of scholars of cinema history that big city life and its version of modernity were exceptional and not typical.[5] Together with European research, work by Robert C. Allen, Kathryn H. Fuller-Seeley and Gregory Waller on film culture in specific rural, small-town and regional town geographies in the United States provides international comparative evidence that film exhibition was construed differently in different places both on a local/regional and national scale.[6] Vanessa Toulmin's work on fairground entertainments and travelling film exhibition in Great Britain, as well as the proceedings of the international conference 'Travelling Cinema in Europe', curated by Martin Loiperdinger, are also of importance to this project.[7] Comparison is essential in tracing the cultural and social correlations and connections that occurred between different locations, in or through the construction of cinematic spaces. It is also essential to the task of unearthing the idiosyncrasies related to Swedish culture, cinema and society.

Based on Rune Waldekranz's impressive panoramic survey of early exhibition patterns in 88 towns and communities spread throughout the country and close-up studies of the first decade of film exhibition in the three regional centres of Örebro, Jönköping and Gävle, the main components of early film exhibition in Sweden can be summed up in a series of propositions.[8] Early film exhibition was concentrated in halls run by voluntary and progressive societies with a largely working-class membership. Film shows were incorporated into existing forms of entertainment that toured the towns and communities throughout the country. Although the forms of exhibition were familiar, the programmes were novel, having films at their core. From 1903 the dominant programme form was composed of a series of films with some kind of documented musical accompaniment. Thus, both in terms of programme and the interior architecture of the exhibition venues, the travelling film show bore considerable similarity to later permanent-site cinemas, available in towns from around 1907.

Touring film exhibitors were a recurring entertainment in towns and communities, increasing in numbers from year to year. From 1904, several exhibitors began keeping their shows in particular towns for longer periods. This involved a transitional kind of practice – neither permanent nor mobile – that competed with yet another transitional exhibition form, a Scandinavian equivalent to the American nickelodeon, in which a shop or café was converted into a cinema that showed continuous half-hour or hour-long programmes for a minimal fee. When permanent-site cinemas opened in towns throughout Sweden around 1907, they could choose to exhibit either a continuous programme or an evening programme that was more closely affiliated to the travelling film show.

The voluntary and progressive societies mentioned above are the key to understanding the spaces of early film exhibition in Sweden and its means of gaining social acceptance. Indeed, they play an essential part in the construction of the Swedish modern welfare state by taking an active part in the transformation of the class-based political system into a democracy through non-revolutionary means.[9] Before considering the centrality of civil society for early film exhibition, a discussion of the emergence of modern forms of civil society in relation to the political climate at hand is required.

Civil society and the people's movements

Grassroots movements of the Free Churches, temperance societies and the labour movement played a central role in the emergence of the Swedish welfare state and the transformation of Swedish society into a democracy. Known as *folkrörelser* ('the people's movements'), these movements were extremely popular, voluntary and self-organized from below. Until recently, Sweden has had a largely homogenous population.[10] In the first few decades of the twentieth century the dominant social stratification was by class, a phenomenon that helps explain the historical centrality of the *folkrörelser*. The extent to which the activities of the Free Churches, the temperance and labour movements should be understood as a functional self-disciplinary project that defused a more radical potential and served to peacefully incorporate the working class into a more or less socialist/liberal yet essentially capitalist economy and culture is a matter of debate among social historians.[11] The historical and political focus on *folkrörelser* at the expense of other associations and societies has of late been called into question.[12] In this chapter, the *folkrörelser* and other organizations belonging to civil society will be explored as social and political forces.[13]

There is no doubt that the three movements were powerful political forces that promoted working-class interests and cultures. Obviously the people's movements also contributed to the late yet peaceful democratization of Sweden and the emergence of a welfare state. The local, regional and national divisions of the movements promoted self-education and practised the protocols of democratic organization, serving as a form of political schooling; some of their members became active politicians in local and national government. On a fundamental ideological level, the three movements shared common ground: the majority of the temperance organizations were permeated by Christian values, and when the Social Democratic Party emerged a large number of its members were also engaged in the temperance movement.[14]

It is important to note, however, that these movements involved a sizeable number of organizations with a national and international reach, with both overlapping and shifting religious, social and political aims. The Free Churches belonged to various Protestant denominations that spread throughout Sweden in the nineteenth century when an evangelical revival broke out, and were united in the common bond of emphasizing personal faith and voluntary religious practice. Younger, working-class women made up a large part of the membership in the Free Churches, yet this group also included people from other social strata.[15] The Free Church movement was especially pervasive in both the regions of Jönköping and Örebro. In Jönköping the Methodist Church was the preferred venue for travelling film exhibitors.[16]

The largest temperance organizations were primarily built on reformist politics concerning the sale and consumption of alcohol, and on Christian values. As within the Free Church movement, social conscientiousness and respectability were important.[17] The temperance societies were also active in the struggle for universal suffrage. In the region of Örebro, the temperance movement regularly distributed three periodicals, one of which was a newspaper, *Nerikes-Tidningen*, equal in size of circulation to the other three regional papers with regular distribution. The ideological identity of *Nerikes-Tidningen* was described by its editor as 'moderately liberal' and as 'nationalistic and pro-universal suffrage'.[18] The temperance movement's adult membership was made up of artisans, skilled workers, civil servants and clerks, with slightly more men than women.[19]

The more radical workers' movement started to organize labour unions and regional societies only in the final decade of the nineteenth century, whereas the Free Churches and the temperance movement had evolved on a large scale in earlier decades. The workers' movement was composed largely of married younger men in the industrial work force who were also qualified workers.[20] If the Free Church and the temperance movements chose a strategy of action that complied with legitimate society, the idea of revolutionary action was an issue for the workers' movement. The general strike in 1909 involved such a threat, as did demands for universal suffrage after 1917. In practice, however, more peaceful, parliamentary methods of political action were preferred.[21] Historian Sven Lundkvist finds the ideals of the provincial workers' societies around the turn of the nineteenth century to be permeated by a 'middle-class' and 'liberal, enterprising ideal' involving a self-help ideology.[22]

Arbetareföreningar were workers' societies founded by industrialists and other prominent bourgeois citizens in a liberal reformist spirit in the second half of the nineteenth century. Their aim was to educate the uneducated adult population and to promote economic self-help strategies such as health insurance and retirement funds.[23] In both Örebro and Gävle these societies played a major part in local film exhibition histories by catering to travelling exhibitors, owning and running the most popular exhibition venue for film in the two towns.

Membership in these civil societies was fluid. It was common to move in and out of an organization or to be a member of more than one society at a time. Until the 1920s the organizations expanded rapidly. While the country's population increased from 5.5 to 5.9 million from 1900 to 1920, adult membership in the three popular movements increased from around 450,000 to around 840,000.[24] Their popularity can be understood as a result of Sweden's late transformation from an agricultural to an industrial, urban economy, which took off in the 1870s. By the 1890s the rate of industrialization, urbanization and modernization was intense by international standards and surpassed most of the other western nations.[25] On a very basic level, the religious, temperance and labour societies offered new forms of social networking at a time and place in which traditional forms of community and family were being disrupted for the newly urbanized working class.

Because these movements were more or less oppositional in their relationship to the state and were at times also persecuted by the authorities, they had to define a space of their own. Building a house where they could congregate in peace became essential. The buildings required loans to fund them. While the religious societies sometimes had wealthy donors among their members or supporters, this was rarely the case with the temperance societies, which relied more heavily on income from bazaars, public parties and travelling entertainers to cover their building costs.[26] Around the turn of the twentieth century, the civil societies provided a nationwide infrastructure that offered travelling showmen a social milieu and places of exhibition that catered to between 200 and 400 people at a time.

Big city versus town exhibition

Most writing on film history pays attention to a different aspect of civil society and its capacity to discipline film exhibition (and indirectly also film production). After permanent-site exhibition venues had opened on an unprecedented scale in the big cities of Gothenburg and Stockholm in 1904–5, *Pedagogiska sällskapet*, a professional society that organized school teachers and pursued a reformist agenda, initiated a public debate about the effects of film exhibition on children. The debate eventually led to the enforcement of national censorship of film exhibition in 1911; the first censorship officials were the members of the teachers' association that had been active in the debate.[27]

The sudden availability of so many permanent exhibition venues for moving pictures was a situation unique to the larger cities of Gothenburg and Stockholm. Before they opened *en masse*, film exhibition was a very marginal form of entertainment in the cities. The popularity of the *Cinématographe Lumière* at the Stockholm Exhibition in 1897 inspired the opening of a few converted shops or cafés where films were screened. Within a couple of years, however, film exhibition in the city was scarce, with infrequent screenings at theatres featuring variety shows.[28] In 1904–5, motion picture venues appeared everywhere, drawing crowds of children to Sunday matinees with the appeal of violent action and crime. The clash between civil society as represented by the teachers' association and this new form of entertainment was inevitable, but the antagonists were reconciled a few years later with the passage of national censorship that not only regulated the content of the screenings according to the standards and tastes of the urban and educated elite, but also drove a number of exchanges and film production companies out of business, leaving AB Svenska Bio, based in Stockholm, to dominate the national film market.[29] Thus, the relations between the government, the industry and the audience became easier to manage and as a result, the Swedish film industry became largely controlled from the capital of Stockholm.

Outside the big city spaces of film exhibition, relations between the local authorities, film exhibition and civil society were very different. Although the extent of local disciplinary actions on exhibition practices has not been thoroughly explored, research on the towns of Örebro, Jönköping and Gävle gives the impression of a disengaged attitude towards local censorship and other regulatory measures, and little concern about restricting advertising in local papers or the effects of film exhibition.[30]

FIGURE 2.1 Man with film projector in a meeting hall in Norrsundet, close to Gävle

Censorship could in theory be carried out by local authorities under a decree dating from 1868.[31] In the large industrial city of Malmö with a population of 90,000 people, the local police hired a schoolteacher as a censor in 1908.[32] In contrast, the authorities in the towns of Örebro, Jönköping and Gävle seem to have remained largely passive. By 1906, it became common practice for the authorities in Örebro to detail a number of conditions in conjunction with granting permission to exhibit film in town, aiming at protecting audiences, especially children, from violent and morally offensive exposure.[33] However, the idea that film exhibition required collective and reformist action was not expressed with conviction in any of the three towns. When a representative from the local teachers' society brought up the question of the negative effects of film on children in a public meeting in Örebro in 1907 with the intent of creating public opinion in favour of censorship, the response was at best tepid. In Jönköping a similar meeting in 1909 generated only a small notice in one of the local papers.[34] By 1908, the lack of interest in film censorship in the provinces was observed by the central authorities, who encouraged local authorities to abide by the rules more stringently.[35]

Although the local press can be described as a political tool for the larger ideological groups, the press in Jönköping, Gävle and Örebro were generally encouraging towards the film business, regardless of their political affiliation. Film exhibition was not considered a major social problem, although the odd violent or sexually provocative film would now and then be subjected to critique by journalists. A closer look at the spaces of film exhibition in Sweden, outside the larger cities, indicates how the business developed through favourable relations with popular sections of civil society in the towns and communities, with the people's movements and the liberal workers' society. In fact, rather than fending off the entertainment and seeking control over its content and function, the two sectors of commercial entertainment and civil society cooperated and nurtured each other, through both economic and social measures.

Spaces of exhibition

One explanation for provincial local authorities' support for the film business can be found in the fact that towns and communities outside the major cities enjoyed more continuous access to this form of entertainment than was available in the big centres, because of the large number of travelling film companies in the country. From Waldekranz's survey, we know that more than 30 touring companies were active in the year 1900, and by 1905 there were as many as 49 touring exhibitors featuring film in their programmes.[36] By comparison, there were 22 travelling circuses and 22 theatre companies on the road in 1900, a higher number than in most years.[37] Observations from Örebro and Gävle show that after 1903 travelling film exhibitors tended to extend their stay in one place.[38] This was most likely a consequence of the increase in the number of films in circulation and the accessibility of more marketable and longer titles. As a result, townspeople became gradually accustomed to the new medium. By the time permanent motion picture venues opened throughout the country in 1907, film exhibition was common and the shock of a 'nickelodeon boom' simply did not occur outside the Big City. In accordance with Vanessa Toulmin's proposal concerning the British film industry, it can be argued that the early years of the moving picture industry in Sweden were shaped not by the exhibition patterns and stimulating cultural effects of the metropolis, but rather by the travelling showmen and their 'other' spaces.[39]

Unlike most of Europe, film exhibition was not a fairground attraction in Sweden before the establishment of permanent venues for cinema. This can partly be explained by the low status of the Swedish fairground. In their analysis of film exhibition at the Nottingham Goose Fair, Marc Jancovich, Lucy Faire and Sarah Stubbings claim that British fairs were in a process of reforming themselves during the period of early film exhibition and consequently appealed to a wider public, including the middle class as well as the lower classes.[40] In sharp contrast, licences for regional markets were being withdrawn one after the other in Sweden, on the grounds that they created unfair competition for the town's permanent retailers and because the fairgrounds were a cause of social disorder.[41]

Travelling exhibitors in Sweden preferred the assembly halls of the popular movements. Waldekranz's survey reveals that close to half the total number of screenings were held in the lodges of the temperance movement. If the survey had also included smaller communities, where the local temperance lodge was the only recreational centre, the temperance movement would appear as an even larger provider of the spaces of early film exhibition. A quarter of the screenings were held in the chapels owned by the Free Churches and between 5 and 10 per cent in venues belonging to the liberal workers' *Arbetareföreningar*. In total, the *folkrörelser* provided between 70 and 80 per cent of the venues for screenings. The remaining shows took place in theatres, secondary schools, in the newly built houses owned by the workers' movement, in riding schools and town halls.[42] With venues belonging to networks of

FIGURE 2.2 Gävle Godtemplars lodge, 1900–1910

FIGURE 2.3 The house of Örebro Arbetareförening, where films were screened between 1897 and 1930: from 1907 under the name of Skandia Biografen

civil society available in larger towns as well as in smaller communities, there was no need to tour the fairgrounds.

Space and programme content

There were clear patterns of correspondence between the programme and the venue of exhibition. In his 1969 survey, Waldekranz suggested that the spaces of film consumption influenced what was shown on the screen.[43] Jancovich, Faire and Stubbings have described this phenomenon in a British context as an effect of films being shown within already established practices of performance, for example, in music halls where film was one act among many in a broad and diverse programme; in town halls where a sporting event could be screened; in churches, where they were used to illustrate sermons; and on the fairgrounds where film was incorporated into a magic lantern- or ghost show. In their words, early cinema represents 'a period of fluidity, in which the meanings of film consumption were less dependent on the film as a text than on the location in which it was consumed'.[44] For Gregory Waller this is a major argument for a local and idiosyncratic kind of film historiography: in his study of film exhibition in Lexington in the American South, Waller claims that different locations of exhibition offered different types of experiences. During the first ten years films were screened at 'the Opera House, Chautauqua Assemblies, churches, summertime theatres, large-scale fairs, and street carnivals', each providing a different and locally specific experience.[45]

Per Vesterlund found similarities in the advertisements for film shows in the local press in Gävle and Lexington, indicating that the cities perhaps shared some idiosyncrasies. At the same time, they were both involved in a process of integration into a globalized media network.[46] As Waldekranz has shown, however, there was no great variety of exhibition sites in Sweden. Film exhibition borrowed advertising practices and some aspects of exhibition from earlier cultural forms. It was common practice at the time for the temperance lodges and other associations to accommodate travelling entertainers, and they also organized their own entertainments. The middle- and lower-class inhabitants and members of the civil societies were already in the habit of attending the halls in their leisure time when film exhibition joined other entertainments on the local programme. Film screenings followed a standard timetable for evening performances, extra performances and matinees, as well as adopting standardized forms of pricing and advertising already in use by touring stage performers.[47]

Jancovich, Faire and Stubbings come to the conclusion that in the period prior to the establishment of the longer and more elaborate feature film, 'films were not considered to provide enough entertainment on their own'.[48] In Sweden, a few touring film companies combined a film screening with a live performance by a variety artist. In the Free Church circuit a lecturer with a slide-show might have film as a highlight in his programme. There were also a few tours that presented joint launches of the phonograph and 'living pictures'. The overwhelming majority of the

FIGURE 2.4 Svea Godtemplars members in front of their lodge near Örebro, 1904

FIGURE 2.5 The Methodist Church in Jönköping where films were frequently screened around 1900

film programmes that have been traced so far, however, showed only films and featured a series of films in their programmes. An evening programme started at 8 p.m., was an hour-and-a-half long and, ideally, offered a rich and varied selection of films that were new to the audience. In the advertisements, the audience that frequented the film shows was promised a diverse spectrum of experiences that ranged from laughter to suspense, from awe to the melodramatic. The programmes exhibited by travelling showmen had a similar structure to those presented by later, permanent-site forms of exhibition.[49]

Spaces of a provincial modernity

Fuller-Seeley and Allen argue that in the US film screenings in the smaller towns eschewed the nickelodeon format of big city film exhibition. In the provinces, film was exhibited in up-scale venues, was socially accepted and, in Charles Musser's description, 'from its outset … drew its audiences from across the working, middle, and elite classes'.[50] As a result, provincial ideals coexisted and competed with the urban ones that came with the film apparatus. Fuller-Seeley has aptly described this as a process in which aspects of urban modernity were allowed to slip into the public discourse of the small town through the back door.[51]

Although there were opportunities for the upper classes to attend film screenings in the smaller towns of Sweden, there is little evidence that film had an appeal to them or indeed that the social elite was in any way bothered with the medium once it had become a common presence in a town. Film was not regularly shown in the up-scale venues, and there was little if any response in public discourse to film exhibition in its first decade. The ease with which film exhibition was accepted in small-town Sweden can be explained by the spaces of its exhibition. Throughout the country a network of social structures and meeting halls had already been constructed by civil society as a reaction and adjustment to modernity and effects of urbanization and industrialization. In this process of adjustment, civil society, in major part represented by the *folkrörelser*, contributed to the transformations of society and to the advancement of modernity. In these circumstances, there was no clear demarcation between a provincial, backward-striving ideology and a modern one. Film fitted nicely into the spaces of the progressive movements. The movements themselves existed as a kind of institution of social mediation between the old and the new – a site of transformation. In the years after the turn of the twentieth century, they were happy to cater to a modern, cheap, socially inclusive attraction that could entice members of the rapidly expanding working and lower middle classes into their halls and into their spaces.

From the 1910s to the 1930s, tensions surfaced between culturally conservative, religious and more progressive camps in the network of *folkrörelser*, yet the popular movements in Swedish civil society continued their prosperous relationship with the commercial entertainment industries, especially with the film and music industries during the post-World War II era. That, however, is another history that is yet to be told.

Notes

1 See for example L. Charney and V.R. Schwartz (eds) *Cinema and the Invention of Modern Life*, Berkeley, Los Angeles, London: University of California Press, 1995; B. Singer, *Melodrama and Modernity: Early sensational cinema and its contexts*, New York: Columbia University Press, 2001. Seminal texts in the theorization of cultural modernity are Charles Baudelaire (1863), *The Painter of Modern Life*; Georg Simmel (1903), *The Metropolis and Mental Life*; Walter Benjamin (1936), *The Work of Art in the Age of Mechanical Reproduction*, all reproduced in V.R. Schwartz and J.M. Przyblyski (eds) *The Nineteenth-Century Visual Culture Reader*, New York: Routledge, 2004.

2 Jameson's concepts are evoked in a discussion concerning the rupture between modernity and postmodernity in A. Friedberg, *Window Shopping: Cinema and the postmodern*, Berkeley: University of California Press, 1993, pp. 168–79.

3 D. Massey, *For Space*, London: Sage, 2005. The advantages of this theory to cinema studies are explored in R.C. Allen, 'The Place of Space in Film Historiography', *Tijdschrift voor Mediageschiedenis* 9, 2006, 15–27.

4 There is only one overview of Swedish film history to date: L. Furhammar, *Filmen i Sverige. En historia i tio kapitel och en fortsättning*, Stockholm: Dialogos/Svenska Filminstitutet, 2003. Pioneering scholars of film have focused their research on the so-called Golden Age between 1917 and 1924, on Great Men and Masterpieces as in the following: B. Idestam-Almquist, *Den svenska filmens drama, Sjöström, Stiller*, Stockholm: Åhlen & söner, 1939; B. Idestam-Almquist, *När filmen kom till Sverige: Charles Magnusson och Svenska Bio*, Stockholm: Norstedt, 1959; G. Werner, *Mauritz Stiller och hans filmer 1912–1916*, Stockholm: Norstedt, 1969; B. Forslund, *Victor Sjöström, hans liv och verk*, Stockholm: Bonnier, 1980. An exception is an unpublished licentiate dissertation on exhibition patterns and films from between 1896 and 1906, divided into the categories of the Big City (Stockholm and Gothenburg) and rural Sweden (constructed through documentation in the local press of 88 towns). R. Waldekranz, *Levande fotografier: film och biograf i Sverige 1896–1906*, Unpubl. Lic. Diss., Stockholm: Stockholm University, 1969.

5 R.C. Allen, 'Race, Region, and Rusticity: relocating U.S. film history', in R. Maltby, M. Stokes and R.C. Allen (eds) *Going to the Movies. Hollywood and the Social Experience of Cinema*, Exeter: Exeter University Press, 2007, pp. 25–44.

6 Allen, 'Race, Region, and Rusticity'; K.H. Fuller-Seeley, *Hollywood in the Neighborhood: historical case studies of local moviegoing*, Berkeley, CA.: University of California Press, 2008; K.H. Fuller, *At the Picture Show: small-town audiences and the creation of movie fan culture*, Washington, D.C.: University of Virginia Press, 1996; G.A. Waller, *Main Street Amusements: movies and commercial entertainment in a southern city 1896–1930*, Washington, D.C.: Smithsonian Institution Press 1995; M. Jancovich and L. Faire with S. Stubbings, *The Place of the Audience: cultural geographies of film consumption*, London: BFI, 2003; see the chapter by D. Biltereyst, Ph. Meers, K. Lotze and L. Van de Vijver in this book.

7 V. Toulmin, 'The importance of the programme in early film presentation', *KINtop* 11, 2002, 19–34; V. Toulmin, 'Telling the tale: The history of the fairground bioscope shows and the showmen who operated them', *Film History* 6, 1994, 219–37; M. Loiperdinger (ed.) 'Travelling cinema in Europe: sources and perspectives', *KINtop* 10, 2008.

8 R. Waldekranz, *Levande fotografier*; Å. Jernudd, *Filmkultur och nöjesliv i Örebro 1897–1908*, Diss., Stockholm: Stockholm University, 2007; Å. Jernudd, 'Före biografens tid: kringresande filmförevisares program 1904–7', in E. Hedling and M. Jönsson (eds) *Välfärdsbilder: svensk film utanför biografen*, Stockholm: Statens ljud-och bildarkiv, 2008; M. Nordström and L. Östvall, *Äventyr i filmbranschen: om entreprenörerna John Johansson och John Ek från Jönköping, deras biografer, distribution och filmproduktion samt nedslag i Jönköpings film-och biografliv 1897–2002*, Jönköping: Jönköpings Läns Museum, 2002; P. Vesterlund, 'Biografkultur i Gävle', in B. Hammar (ed.) *Medierade offentligheter och identitet*, Gävle: Högskolan i Gävle, 2006. The towns Örebro, Jönköping and Gävle all belong to the densely populated southern half of the country and were regional centres of about the same size, attracting hordes of people to the newly established industries around the turn of the twentieth century. In 1892, Örebro had a population of 14,893, Jönköping 19,902 and Gävle 25,008. By 1912 the populations had increased in Örebro to 32,075, Jönköping 27,864 and Gävle 35,838. In a demographic survey conducted in 1912 the three towns were grouped among the 30 larger centres in the country with a population of more than 10,000. The sum total for the 30 towns (which included the relative giants Stockholm and Gothenburg) were in 1892: 922,348 and in 1912: 1,434,232. This corresponded to 21 per cent of the population of the nation as a whole. Although the circumstances in Örebro, Jönköping and Gävle do not reflect the historical situation that was characteristic of the majority of the population in rural Sweden, a study of them sheds light on the spaces of travelling film exhibition that toured towns as well as smaller communities outside of the big cities. 'Befolkningsrörelsen år 1912. Folkmängden och dess förändringar', in *Svensk offentlig statistik*, Stockholm: Statistiska centralbyrån, 1915.

9 What is commonly referred to as the Swedish 'third way model' of society lasted roughly from the 1930s to the mid-1980s and is distinguished by two characteristics: first, the long, only partially disrupted rule of the Social Democratic Party and second, its close affiliation with the labour unions. The bond between the labour unions and the Social Democratic Party made labour market stability a top priority in national politics. The welfare state that was conceived under these conditions ruled under the auspices of an ideology that sought modernization through a consensus-oriented politics stressing national unity and social harmony. However, this economic model is largely a result of lessons learned from the instability in Europe during the interwar period. There is no simple progressive relation between the pre-war, inter- and post-war periods. L. Schön, *En modern svensk ekonomisk historia: tillväxt och omvandling under två sekel*, Stockholm: SNS förlag, 2007, pp. 401–2.

10 In 1999 the Swedish government officially recognized five minorities: the Jewish population, the Romani, the aboriginal Sami population in the far North, Swedish-Finns and *Tornedalingar*. http://www.sweden.gov.se/content/1/c6/08/56/33/2fe839be.pdf.

11 S. Lundkvist, *Folkrörelserna i det svenska samhället 1850–1920*, Uppsala: Uppsala University, 1977; B. Horgby, *Egensinne och skötsamhet: arbetarkulturen i Norrköping 1850–1940*, Stockholm: Carlsson, 1993; R. Ambjörnsson, *Den skötsamme arbetaren: idéer och ideal i ett norrländskt sågverkssamhälle 1880–1930*, Stockholm: Carlsson, 1988.

12 The term 'civil society' has been used to reconsider 'the Swedish model' of social, political and economic life in the wake of globalization, regionalization and the failure to sustain traditional social democratic ideals of welfare and democracy. L. Trägårdh (ed.) *State and Civil Society in Northern Europe. The Swedish Model Reconsidered*, New York: Berghahn Books, 2007; E. Amnå, 'Still a Trustworthy Ally? Civil Society and the Transformation of Scandinavian Democracy', *Journal of Civil Society* 2, 2006, 1–20; E. Amnå (ed.) *Civilsamhället: några forskningsfrågor*, Stockholm: Riksbankens jubileumsfond/Gidlund, 2005.

13 'Civil society' is a useful term in this context in that it accentuates how, in a small town perspective, social relations emanate from the life conditions associated with the advance of modernity. In this history the concept of civil society will be used to designate voluntary associations and forms of social deliberation that constitute 'the third sector', independent of the state and the commercial sector. It is used to include multifarious social and political societies that were of importance to the creation of the spaces of early cinema in Sweden. In using the term 'civil society' instead of *folkrörelser* it can be argued that the political importance of the people's movements is de-emphasized. However, it can also be argued that the term 'civil society' offers a more critical perspective loyal to the civilians rather than to the state. E. Amnå, 'Scenöppning, scenvridning, scenför-ändring: en introduktion', in E. Amnå (ed.) *Civilsamhället: några forskningsfrågor*, Stockholm: Riksbankens jubileumsfond/Gidlund, 2005, p. 17. Another reason for adopting the term is that it can include other kinds of voluntary associations that are not included in the term *folkrörelser,* but were important to the spaces of early film exhibition in Sweden.

14 T. Frängsmyr, *Svensk idéhistoria. Bildning och vetenskap under tusen år, del II 1809–2000*, Stockholm: Natur och kultur, 2000, p. 186; Lundqvist, *Folkrörelserna i det svenska samhället*, pp. 71–75.

15 Lundqvist, *Folkrörelserna i det svenska samhället*, pp. 47–50, 58–59, 126–29.

16 Nordström and Östvall, *Äventyr i filmbranschen*, p. 45.

17 R. Ambjörnsson, *Den skötsamme arbetaren: idéer och ideal i ett norrländskt sågverkssamhälle 1880–1930*, Stockholm: Carlsson, 1988; Frängsmyr, *Svensk idéhistoria*, p. 186.

18 *Nerikes-Tidningen*, 21 January 1899.

19 Lundqvist, *Folkrörelserna i det svenska samhället*, pp. 127–28.

20 Ibid., p. 128.

21 Ibid., p. 165.

22 Ibid., p. 194; Frängsmyr, *Svensk idéhistoria*, pp. 187–90.

23 Frängsmyr, *Svensk idéhistoria*, p. 187.

24 Lundqvist, *Folkrörelserna i det svenska samhället*, pp. 64–67.

25 Schön, *En modern svensk ekonomisk historia*, p. 220.

26 J.-B. Schnell, 'Folkrörelsernas byggnader', in T. Hall and K. Dunér (eds) *Svenska Hus: Landsbygdens arkitektur – från bondesamhälle till industrialism*, Stockholm: Carlsson/Riksantikvarieämbetet/Sverige Radio, 1999, pp. 197–98. The workers' movement also contended for a house of their own. These came to be known as *Folkets Hus*, 'the peo-ple's house'. They were used for political meetings, but also for dances, film and theatre shows. In 1952, there were 955 *Folkets Hus* registered throughout the country and of these, 500 regularly exhibited film. Initially, it was difficult to fund the building of a house whereas it was easier to open a fenced-in leisure park where social activities such as Tivoli, performances and dances, could be organized and a small entrance fee could be charged. These are known as *Folkets Park*, 'the people's park'. From the 1930s to the early 1970s they provided seasonal summer entertainment throughout the country, and in conjunction with the radio, theatre and film industry they presented a steady stream of

national and international entertainers. At the most, 250 parks joined the national *Folkets Park* organization. http://www.fhp.nu.

27 Furhammar, *Filmen i Sverige,* pp. 39–41. In Waldekranz's history, the articles in the national press on the subject, most of which were written by members of *Pedagogiska Sällskapet,* are equated with 'public opinion'. Waldekranz, *Levande fotografier,* pp. 267–72.

28 Waldekranz, *Levande fotografier,* pp. 24–113, 247–66, 280. Film historian Jan Olsson has suggested that one of the pioneering film exhibitors in Stockholm closed his permanent-site cinema which had been in operation for a few months following the Stockholm Exhibition in 1897 because of lack of films. J. Olsson (1999), 'Exchange and Exhibition Practices: notes on the Swedish market in the transitional era', in J. Fullerton and J. Olsson (eds) *Nordic Explorations: Film Before 1930,* London: Indiana University Press, 1999, p. 147.

29 J. Olsson, 'Svart på vitt: film, makt och censur', *Aura. Film Studies Journal* 1, 1995, 14–46; J. Olsson, 'Exchange and Exhibition Practices'.

30 Jernudd, *Filmkultur och nöjesliv i Örebro;* Nordström and Östvall, *Äventyr i filmbranschen;* Vesterlund, 'Biografkultur i Gävle'.

31 E. Skoglund, *Filmcensuren,* Stockholm: PAN/Norstedt, 1971, pp. 9–14.

32 J. Olsson, *Sensationer från en bakgård: Frans Lundberg som biografägare och filmproducent i Malmö och Köpenhamn,* Lund, Stockholm: Symposion Bokförlag, 1988, pp. 46–49; 'Befolkningsrörelsen år 1912'.

33 Jernudd, *Filmkultur och nöjesliv i Örebro,* p. 135.

34 Nordström and Östvall, *Äventyr i filmbranschen,* p. 126.

35 Jernudd, *Filmkultur och nöjesliv i Örebro,* pp. 149–54.

36 Waldekranz, *Levande fotografier,* pp. 117–19.

37 Ibid., p. 148.

38 Jernudd, *Filmkultur och nöjesliv i Örebro,* pp. 129–33; Vesterlund, 'Biografkultur i Gävle', p. 78. That this tendency was not observed in the study on Jönköping can be explained by the fact that the venues of exhibition preferred by travelling exhibitors were different and did not offer themselves easily to being booked on a long-term basis. In Jönköping, the venues of the Free Churches dominated exhibition. Nordström and Östvall, *Äventyr i filmbranschen,* p. 45.

39 Toulmin, 'Telling the tale', pp. 219–37.

40 Jancovich *et al.* also refer to changes regarding working class culture in Britain at the time. British working-class culture is described as conservative in that inequalities of class were apprehended as inevitable. Furthermore, reformist ideas were rejected in favour of a positive attitude towards commercial amusements. As a consequence, the fairground was not a place of carnival and it was not associated with anti-bourgeois subversion. Jancovich, Faire and Stubbings, *The Place of the Audience,* pp. 54–56. For an overview of research on film exhibition in the European fairground traditions, see Loiperdinger, *Travelling Cinema in Europe.*

41 Jernudd, *Filmkultur och nöjesliv i Örebro,* pp. 78–80.

42 The first survey documents the initial screenings of film in the chosen places and the second survey covers the year 1900. The survey was conducted through a search for advertisements in the local press covering 88 towns and communities. Waldekranz, *Levande fotografier,* pp. 216–21.

43 Waldekranz, *Levande fotografier,* p. 216.

44 Jancovich, Faire and Stubbings, *The Place of the Audience,* p. 37.

45 Waller, *Main Street Amusements,* pp. 37, 258.

46 Vesterlund, 'Biografkultur i Gävle', p. 95.

47 Jernudd, *Filmkultur och nöjesliv i Örebro,* pp. 110–12.

48 Jancovich, Faire and Stubbings, *The Place of the Audience,* p. 63.

49 Waldekranz, *Levande fotografier,* pp. 163–214; Jernudd, *Filmkultur och nöjesliv i Örebro,* pp. 87–96; Jernudd, 'Före biografens tid'. The travelling film exhibitors and their

programmes that surface in the studies on Gävle and Jönköping are all familiar in the Örebro material.

50 Ch. Musser, *The Emergence of Cinema*, Berkeley: University of California Press, 1990, p. 183.

51 K.H. Fuller-Seeley, 'Modernity for Small Town Tastes: Movies at the 1907 Cooperstown, New York, Centennial', in R. Maltby, D. Biltereyst and Ph. Meers (eds) *Explorations in New Cinema History: Approaches and Case Studies*, Malden, MA: Wiley-Blackwell, 2011.

3

MOVIEGOING UNDER MILITARY OCCUPATION

Düsseldorf, 1919–25

Frank Kessler and Sabine Lenk

To look at the history of cinema through the experience of moviegoing involves taking the particular circumstances in which audiences encounter moving pictures as a starting point. The frequency of local or regional studies and other forms of 'micro-history' is a consequence of this form of investigation. As a rule, however, cinema historians look at situations that represent a degree of 'normality' and permit extra-polations, or generalizations to a broader level. Analyses of moviegoing under more exceptional circumstances such as political crises, conflicts or changes are less common: such moments may appear too atypical, too much out of the ordinary to reveal anything that can be taken beyond the anecdotal or incidental. On the other hand, such situations may also shed new light on the 'normal' state of things by simply marking the possibility of a difference.

The particular case discussed here, the conditions under which the inhabitants of the German city of Düsseldorf went to see films during the period of Allied occupation after the First World War, is of interest in several ways. First, it concerns a situation in which going to the movies, or going to a particular theatre or show, could be linked to issues of patriotism and political loyalties. Second, cinema owners in these cir-cumstances found themselves navigating between their commercial interests and their obligations to the community, between constraints imposed by foreign authorities and the demands of their audiences. Finally, during this period the conditions of moviegoing in Düsseldorf and the occupied Rhine territories were at least partially different from the rest of the country, and this fact tends to be neglected by tradi-tional film histories, which generally seem to presuppose a homogeneous landscape of cinema in Germany during those years.[1]

Several factors affected the way in which military occupation had an impact on the cinemagoing experience of Düsseldorf audiences. At the highest organizational level, a number of regulations promulgated by the Inter-Allied High Commission for the occupied Rhine territories concerned cinema directly or indirectly. After discussing the general framework that these regulations created for municipalities, exhibitors and audiences, we will examine how they were put into effect in Düsseldorf, and the kinds of conflicts that arose from their implementation. In the third section of this chapter we then consider what films were available in Düsseldorf during this period in order to see to what extent the occupation had an impact on film programmes and film supply. In doing so we want to address the way in which political authorities, exhibitors and audiences had to negotiate the different forces and interests that acted upon the experience of moviegoing. We are particularly interested to see at what levels this experience was actually affected by the political situation.[2]

Entertainment or propaganda? The Inter-Allied High Commission and the cinema

Although the Armistice put an end to fighting on 11 November 1918, the First World War did not end then. The Versailles peace negotiations still had to be conducted, and even after their conclusion and the ratification of the treaty Allied forces continued to occupy parts of German territory, including large parts of the Rhineland. This military presence maintained pressure on the Reich's government. When Germany was unable (or unwilling, as some of the Allies saw it) to fulfil its obligations, French and Belgian forces extended the occupied zone into the industrial Ruhr area. The occupation of Düsseldorf did not end until 25 August 1925, and the last Allied troops did not leave the Rhineland until 1930.

The regulations concerning the military occupation of the territories of the Rhine were stipulated in an agreement between Germany and the Allied countries of the United States, Belgium, Great Britain and France, signed on 28 June 1919 in Versailles. These had an impact on the sovereignty of the Reich in a number of areas. Article 3 (c), for instance, declared that: 'The German courts shall continue to exercise civil and criminal jurisdiction subject to the exceptions contained in paragraphs (d) and (e).'[3] The exceptions referred to persons employed by or in the services of Allied troops as well as persons committing offences against Allied troops, limiting German jurisdiction over all matters involving the occupying forces, including cases in which Allied soldiers clearly violated existing laws. Article 5 stated:

> The civil administration of the provinces (*Provinzen*), Government departments (*Regierungsbezirke*), Urban Circles (*Stadtkreise*), Rural Circles (*Landkreise*) and Communes (*Gemeinde*) shall remain in the hands of the German authorities, and the civil administration of these areas shall continue under German law and under the authority of the Central German Government, except in so far as it may be necessary for the High Commission [...] to adapt that administration to the needs and circumstances of military occupation. [...][4]

All legal and other regulations decided upon by the German government had to be approved by the High Commission before they could be applied to the territories of the Rhine. In addition, the High Commission issued ordinances that defined the various types of facilities, forms of logistical support and housing arrangements that the German authorities had to provide for the Allied troops.[5] These included not only public buildings such as schools and barracks, but also private homes where Allied officers and soldiers took their quarters. During and after the occupation, the German authorities offered compensation to all those who suffered any kind of financial loss or damage caused by the foreign troops' presence.[6]

Like most aspects of everyday life, cinema was affected by these regulations. The authorities in the occupied parts of the country had to provide entertainment for the troops, and article 8 (a) of the agreement stipulated that:

> The German Government shall undertake, moreover, to place at the disposal of the Allied and Associated troops and to maintain in good state of repair all the military establishments required for the said troops. [...] These shall include [...] also theatre and cinema premises [...].[7]

In practice, this meant that movie theatres, dance halls, restaurants and even church premises were regularly seized by the military authorities for a few hours, or sometimes for days or months, in order to organize film shows and other entertainments for the troops. For instance, a cinema owner in Benrath, a small town south of Düsseldorf, placed advertisements in the local paper to announce that 'unless my theatre is occupied by troops, there will be cinematographic shows on Saturday the 21st and Sunday the 22nd of this month'.[8] Quite often the military authorities also granted German civilians access to film shows organized for the soldiers. In most cases, they had to pay for their tickets, although prices were generally lower than in ordinary cinemas. In some cities, including Essen, there were also shows, usually consisting of propaganda films, that Germans could attend for free.[9] In other cities the military authorities did not organize their own shows, but made sure that the soldiers could benefit from the regular programmes offered by the local theatres. A report compiled by the authorities in Datteln asserts that members of the Allied forces only paid half price in cinemas, that the best seats for officers were no more than 5 francs maximum, and 1.50–3 francs for common soldiers, depending on the quality of the theatre.[10]

These regulations meant that from time to time the local population in the occupied zones could not get into their neighbourhood cinema, because it was seized by the military authorities. On other occasions, they might have had to decide whether or not to attend a show that clearly was organized for propaganda purposes, or watch a programme at a reduced price in a cinema operated by the military, at the risk of being considered unpatriotic by their fellow citizens. When cinemas were run commercially under military control, it might even be an economic necessity for the exhibitor to pull in the local crowds, in spite of protests voiced within the community. In other words, for a movie fan in an occupied area the desire to go and see a movie may have provoked private or public conflicts about national loyalty and patriotism.

As for the movies that could be screened, the High Command issued a number of measures that put films and film shows under their control.[11] Ordinance no. 3

concerned the press as well as all other types of publication, and gave Allied authorities the right to censor or ban whatever they felt was detrimental to them or their troops:

> All newspapers, pamphlets or publications, all printed matter, all productions obtained by mechanical or chemical methods, intended for public distribution, all pictures with or without words, all music with words or explanations, and all cinematograph films of a nature that prejudices public order or endangers the security or the dignity of the High Commission or of the troops of occupation are forbidden and if published may be seized by order of the High Commission […].[12]

In addition, stage performances, moving picture shows, pantomimes, lectures, concerts, speeches or other public presentations 'of a nature that prejudices public order or affects the security or dignity of the High Commission or of the troops of Occupation' were banned.[13] A German commentator on these regulations remarked with some sarcasm that the Allied authorities had introduced a new type of offence, the 'threat to public order posed by the press'.[14] Film was subject to censorship at two levels: as text and as performance, and this dual-level control may explain why the film *Der Rhein* (*The Rhine*), consisting mainly of pictures of towns and landscapes, was considered as endangering the dignity of the occupation forces and an insult to France: a simple screening of the film might be perceived as celebrating the Rhine as a German river and inherently an act of opposition to the French.[15]

Not surprisingly, films that even indirectly represented states of occupation were banned, as were historical epics dealing with confrontations between Germany and countries belonging to the Allied forces, films that might create prejudice against the Allied nations, or films featuring demonstrations of military force that might provoke nationalist feelings in the German audience. The reasons why the screening of films such as *Deutsche Helden in schwerer Zeit* (*German Heroes in Hard Times*, Kurt Blach-nitzky, Feindt, Germany 1924), *Mutter Donau, Vater Rhein* (*Mother Danube, Father Rhine*, Heinrich Reichmann, Alfas, Germany 1924), *Die Wacht am Rhein* (*Watch on the Rhine*, Helene Lackner, Koop, Germany 1926) were obvious.[16] *Der Stier von Olivera* (*The Bull from Olivera*, Erich Schönfelder, Ufa, Germany 1920/21), set during the Napoleonic Wars and depicting the French as occupants while glorifying Spanish resistance, was banned, because it 'insulted France' and was accordingly considered a danger to the dignity of the occupation forces.[17]

A negative representation of the French may have similarly led to the prohibition of *Lederstrumpf, 2. Teil: Der Letzte der Mohikaner* (*Leatherstocking Part 2: The Last of the Mohicans*, Arthur Wellin, Luna-Film, Germany 1920), a German adaptation of James Fenimore Cooper's famous novel, although in this case the ban was lifted after some changes had been made.[18] *Reveille: Das große Wecken* (*Reveille: The Great Awakening*, Fritz Kaufmann, Omnium Film-Co., Germany 1920) was also initially banned and later approved, 'provided that the representations of military parades are omitted', the Omnium Film-Co itself having declared that these scenes were not material to the film's subject.[19]

Military parade scenes as such could thus be considered problematic by the Allied authorities, even when they appeared in an otherwise inoffensive context, as apparently was the case with *Reveille*. This example indicates the considerable degree of suspicion guiding the High Commission's censorship efforts. For moviegoers in the occupied areas this meant that the range of films available partially differed from what was screened elsewhere in Germany, and at the same time they may have been particularly sensitive for whatever might give the impression of carrying a 'hidden message', or something that the Allied authorities might suspect of being one.

In some parts of the occupied territories the censorship decrees seem not to have been applied as rigorously as elsewhere. The report on the period of occupation in the town of Derne, for instance, simply stated: 'A censorship of films or stage plays did not occur.'[20] The criteria used by the High Commission to ban a film were also sometimes obscure. In occupied Benrath, for instance, in December 1921 the Schloss-Theater showed the film *Der Doppelmord von Sarajevo* (*The Double Murder at Sarajevo*, Rolf Randolf, Rolf Randolf-Film, Germany 1919/20). The advertisement suggested that this film dealt with the question of who was guilty of having provoked the First World War, and there was no doubt that Serbia was blamed.[21] It is surprising that the High Commission did not censor this film, given that it obviously took a position on the causes of the First World War at odds with the Allied view that Austria and Germany were unquestionably responsible for the conflict. One explanation might be that questions of a more political nature were seen as secondary to direct or indirect references to the state of occupation and possible damage to the dignity of the Allies and their troops.

On the other hand, film programmes in the occupied territories may have differed from the rest of Germany because in certain circumstances films originating from the Allied countries could escape German control. The High Commission's Ordinance no. 74 dated February 1921 stipulated the following:

> The application in the Occupied Territories of the German Ordinances of 16th January, 1917 and 22nd March, 1920 regarding the regulation of imports, and of the German law of 12th May, 1920, regarding the censorship of films, is suspended as far as it concerns films which originate in any of the countries or possessions of the Powers participating in the Occupation and of which the explanatory words are in the German language and one or more of the languages used in the occupying Armies of such Powers.[22]

In other words, imported films from Allied countries with intertitles in German and in one or more of the occupation troops' languages, were exempt from German regulations and censorship. In the first instance, of course, this was meant to prevent German authorities from banning films they might consider foreign propaganda. This could also open possibilities for producers and distributors from Allied countries to circumvent German regulations and censorship simply by using multilingual intertitles. German officials suspected that although German exhibitors complained that bilingual films were less successful with their audience, they were actually benefiting from this situation by sidestepping German censorship. A report to the German Commissioner for the occupied territories of the Rhine (*Reichskommissar für die besetzten rheinischen Gebiete*) from 21 March 1921 expressed some concern about this issue, as well as a profound distrust of German cinema owners:

> A licence to show bilingual foreign films would have incalculable consequences. The fact that several owners of some of the larger moving picture theatres have declared French films not to be attractive enough to be shown profitably in the occupied territories does not, in my opinion, draw the right conclusions in this matter. In all probability, the theatre owners who have been heard declared this in their own interest, so that they would get the opportunity to screen uncensored foreign films. The quality and effect of such films on the population is indeed incalculable. The statement that all foreign films have little effect on the taste of the people living in the occupied territories, and that showing them would not be profitable still needs to be proven. There is proof, however, that some moving picture theatres show films of French origin quite frequently, or even exclusively, and one can surely not presume that this is disadvantageous for them.[23]

German officials seem to have been in favour of the Allied troops seizing theatres temporarily or permanently, provided that German audiences were not admitted to screenings of films that had not been submitted to German censorship, but the High Commission refused to implement this.[24] Eventually, in an amendment to Ordinance 74 from October 1922, the High Commission opened at least some possibilities for German authorities to intervene:

> Whenever the German Authorities shall consider that a film specified in Article 1 shall be such as to be injurious to morals they shall inform the Representative of the High Commission in the Kreis, who, if he shall be in agreement, may permit the German Authorities immediately to forbid either the exhibition of the film wholly or in part or the admission of young persons under 18 years of age to the exhibition of such films in the Kreis concerned. He shall report his action to the High Commission who may maintain or cancel the decision of the Representative and may order that the same measures be applied to such a film, throughout the Occupied Territories.[25]

This at least provided a way to prevent the presentation of foreign films considered unsuitable for German audiences, albeit on moral grounds only, but more often than not, German officials had to accept the fact that part of what was shown on the screens was outside their jurisdiction.

In general, therefore, the cinemagoing experience in the occupied territories could differ from the one elsewhere in Germany both at the level of the show itself and with regard to the films screened. Movie theatres (or other locations fulfilling this function) could be temporarily seized by the troops and thus be inaccessible to other patrons, or else foreign soldiers attended regular screenings, mixing with the locals. In some cases special shows were organized addressing German audiences as well as Allied soldiers, but visiting such an event might be interpreted as unpatriotic. The availability of films could be reduced as a result of bans pronounced by the High Commission, but it could also include imported material that had bypassed German censorship regulations. At this level of regulation, the interests of Allied and German authorities in fact conflicted, mainly in regard to the scope of control exercised by them. For the military administration, the most important issue was to avoid any kind of resistance to the occupation, and their main concern was to ban films and other types of cultural products they feared might spark German patriotism. For the Germans, on the other hand, the question of censorship involved first and foremost control of the films' morals. Given the possibility of circumventing German rules, the authorities suspected that exhibitors benefited from the situation.[26] In this case, political and commercial interests clashed along lines only indirectly related to those of nationality.

Conflicts, compromises and competition: film exhibition in occupied Düsseldorf

The Rhine divides Düsseldorf, with the main part of the city being on the east bank of the river, while on the western bank there are two districts, Oberkassel and Heerdt. Düsseldorf, today the capital of North Rhine-Westphalia, was and is the most important economic centre between Cologne and the Ruhr area.[27]

On 4 December 1918, less than a month after the Armistice, Belgian troops occupied the districts of Oberkassel and Heerdt. The bridge connecting these districts with the east bank, the Oberkasseler Brücke, was blocked, severely restricting communication between the two parts of Düsseldorf and cutting off the inhabitants of the western bank from the centre of the city. Subsequently, whenever the Allied nations decided on measures in response to what they perceived as Germany's unwillingness to fulfil her obligations under the Treaty of Versailles, they extended the occupied zone, most famously with the invasion of the Ruhr area in 1923. In March 1921 the French army crossed the Rhine and occupied the remaining parts of Düsseldorf, where they stayed until August 1925.[28] Although restrictions on the circulation between the two parts of Düsseldorf were later lifted or at least considerably reduced, the military presence weighed heavily upon the city, and especially on the left bank of the Rhine. There were 3,000 foreign troops stationed in Oberkassel alone on a regular basis, and that number rose to 10,000 in times of heightened political tensions. These soldiers were quartered in schools, restaurants or factory halls, while officers and their families were generally housed in private homes. Facilities of various kinds, including entertainment venues, had to be made available to the troops, or else were seized by the military authorities, and this often had negative financial effects for the proprietors.

In the main part of Düsseldorf on the right bank of the Rhine, the French apparently decided not to operate a military cinema themselves, instead making sure that their troops could benefit from the entertainments offered in town. An ordinance of 14 May 1921 stipulated that the municipal theatre and the largest variety theatre in town, the Apollo, had to reserve a considerable number of seats for officers and common soldiers at reduced prices.[29] In a letter dated 26 May 1921, General Magni, the commanding officer, sent an urgent request to the Lord Mayor of Düsseldorf, ordering him to immediately provide the military authorities with a list of all the cinemas in town, specifying the exact address, the number of seats and the type of theatre.[30] A draft report on the period of occupation asserts that by 'agreement' between the Representative of the High Commission and a delegation of Düsseldorf cinema owners, prices for foreign troops at Düsseldorf's four major cinemas, the Residenz-Lichtspiele, the Alhambra-Lichtspiele, the Ufa-Lichtspiele and the Asta Nielsen, were reduced to between one quarter and one third of the regular entrance fee.[31] The owners, however, maintained that the 'agreement' was in fact an order, but that in spite of their efforts they could not succeed in getting this in writing from the High Commission.[32] The absence of a written order had consequences for the exhibitors: since there was no proof that they had been officially compelled to reduce their prices, they could not claim compensation for their financial loss.

One might speculate that the German officials' negative attitude towards the cinema owners was inspired by the suspicion that the exhibitors benefited from showing films with bilingual intertitles, as this allowed them not only to sidestep German censorship regulations, but also to increase their audience numbers, even though the foreign soldiers only paid half-price (during the war the German soldiers had actually paid the same reduced fee). This issue of national loyalty came to the fore when in June 1923, Fritz Genandt, the manager of the Residenz-Lichtspiele, had to defend himself before the Lord Mayor against an accusation that he had engaged in 'shameful conduct' in apologizing to his audience for screening a print of *Der rote Reiter* (*The Red Horseman*) that only had German, and not bilingual intertitles.[33] Genandt justified himself by declaring that the Residenz, like every other cinema in town, was forced to show films in bilingual versions, on the basis of a verbal order from the military authorities.[34] Although this may have been an isolated incident, it clearly shows that there were people who took offence when public screenings explicitly addressed the foreign soldiers as well as the German audience, even though the theatres had no choice in this matter.

Although audiences were continually reminded of the occupied status of their city by the bilingual intertitles, it seems from the available documentation that cinema-going was not especially restricted by the military authorities in Düsseldorf. In the draft report about the occupation there is no mention of any censorship measures taken against individual films, although other documents suggest that at least in the beginning, the military authorities paid attention to what was screened.[35] There also is no evidence that cinemas were used for propaganda screenings, although the report suggests that in 1923 the French projected films onto a screen placed inside the window of the French reading room provided 'For Germans only' on Düsseldorf's most elegant boulevard, Königsallee, so that they 'could be watched from the street. The groups of viewers standing there did of course cause traffic jams … In order to attract people passing by, comedies were shown in-between the propaganda films (French landscapes and cities, etc.).' The report also claimed that the French authorities would not have tolerated any attempt by the German police to disperse the crowds.[36] It also indicates that the German authorities considered travelogues of French landscapes to be propaganda material, in the same way that the Inter-Allied High Commission regarded *Der Rhein*.

On the West bank of the Rhine, Oberkassel was occupied by Belgian troops from December 1918, and initially the district was more or less cut off from the city centre. Here the authorities opted for a special screening facility for their troops, and this case illustrates the kind of conflicts that arose from the regulations stipulated in the agreement between Germany and the Allies about the occupation. The first location to be used by the Belgian military to organize film shows and other types of entertainment for their soldiers was, strangely enough, a Protestant church located in Arnulfstrasse.[37] In the summer of 1919 the authorities seized a large room in the basement, where the troops could play billiards, listen to music, and watch films and other kinds of shows. For the German population, and the Protestant community in particular, this was close to a desecration of the building, especially as this specific room had formerly served as a space for religious education. People were further upset when the military authorities issued a decree permitting Germans to attend entertainments there. After protestations from the church's minister and congregation and negotiations with the mayor which failed to identify an alternative venue, Germans were again excluded in December 1919.[38] The Allied forces seem to have regarded German attendance at film shows for the troops as an issue of some importance. This may have been in part an economic matter, since non-army visitors would have contributed to financing the troops' entertainment, but it is also quite possible that the military authorities thought that such screenings could have a positive effect on the Germans' perception of their occupiers. Eventually, in March 1920, a large room in the Gaststätte Bollig that normally had been used as a ballroom by various clubs and associations was transformed into a screening facility for the Belgian troops.[39] The municipality had to cover the monthly rent of 3,000 Marks, while the military authorities organized the shows and their advertising, supplied the films and kept the entrance fees paid by the civilians who attended the shows. The municipality was not allowed to apply an amusement tax to the tickets sold to civilians. Since the amusement tax made up 60 per cent of the ticket price, this also allowed the manager of the military cinema to charge lower prices than his civilian competitors. Oberkassel moviegoers benefited financially from the occupation, while the other exhibitors endured what they must have considered unfair competition.

The Belgian troops also made use of a movie theatre in Belsenstraße.[40] The cinema was a light, wooden construction, built as a temporary installation, and opened in January 1919 because the owner hoped to take advantage of the fact that the residents of Oberkassel were more or less cut off from the public entertainment facilities in the city centre. In January 1920, the manager Georg Wieczorek agreed to let the Belgian military use the cinema, but then objected to the requisition of the theatre for four days a week on the grounds that the Belgian authorities planned to admit Germans to the shows at reduced ticket prices, effectively setting themselves up as a competitor to his business within his own cinema.[41] Wieczorek may well have hoped to hand over the business entirely and receive a handsome compensation in exchange, but when this did not eventuate, he complained that the presence of Belgian soldiers kept away the more sophisticated public, who refused to mingle with the foreign soldiers, and asked the municipal authorities to take over the cinema and compensate him. Subsequently he and the new owner must have tried to make things difficult for the Belgians, since the Belgian Commanding Officer complained in June 1920 about 'unpleasant quarrels', threatening to close the venue. By October 1921, the cinema had been closed.[42]

The history of military film shows in Oberkassel, if they continued, cannot be reconstructed from the available documents. Possibly, with the restrictions being gradually lifted, soldiers crossed the bridge and went into the centre of Düsseldorf instead. However, even this limited case study allows us to draw at least some provisional conclusions. The fact that facilities had to be made available to foreign troops not only meant that local businessmen – cinema managers, but also owners of bars, dance halls or other entertainment venues – lost money because they could not use their premises as they wished. Alternatively, they had to admit foreign soldiers at reduced prices, which kept away many of the usual German patrons.[43] The military presence also meant that other institutions, including churches, had to deal with requisitions. The fact that for economic as well as propaganda reasons, the occupation forces wanted to admit Germans to the shows created additional problems for local cinemas, as the military exhibitors could offer considerably lower rates, because of their exemption from the amusement tax. The incidents discussed here also show that in the everyday practice of moviegoing, audiences were regularly confronted by the fact that they lived in an occupied zone. The presence of foreign soldiers in local cinemas, special programmes, and the fact that they often saw bilingual intertitles, were constant reminders of this. More importantly, the very act of going to see a movie at a specific place at a specific time could raise issues of patriotism and national loyalty.

Business as usual? Film programmes in Düsseldorf[44]

In the week of 11 November 1918, when the Armistice had put an end to the battles of the First World War, the Düsseldorf cinemas showed mainly romantic melodramas and sensational detective stories featuring popular actors such as Henny Porten, Pola Negri and Max Landa.[45] The political situation seemed not to have any visible effect on the films on offer in the city. Not even the occupation of Heerdt and Oberkassel about a month later affected the programmes, with one exception: the UT-Lichtspiele, owned by Fritz Genandt, announced on Friday 15 November that singers from the Düsseldorf Stadttheater would perform *Das Volkslied. Der erste Tag der Republik* (*The Folk Song. The First Day of the Republic*), and this continued to be part of their programme for at least a week.[46] Either by coincidence or as a political statement, the UT-Lichtspiele had members of the municipal theatre company sing a

Republikanisches Wiegenlied (Republican Lullaby) two days after the occupation of the left bank by the Belgians.[47] The fact that local artists were asked to sing songs highlighting republican ideas as a regular part of a movie theatre programme was remarkable in times of turmoil and revolution, although the films shown at the UT-Lichtspiele do not otherwise betray any particular political sympathies.

Interestingly, during the week preceding the occupation of the main part of Düsseldorf by British, Belgian and French troops on 8 March 1921, no advertisements for film programmes appeared in the *Düsseldorfer Nachrichten*. In the following weeks, however, cinemas announced their programmes regularly, before further periods without advertisements in April and May. It is not clear whether this has anything to do with the occupation, nor does this mean that there were no film shows at those dates. In any event, if there were any disruptions due to interventions by the military, they appear to have been temporary.

A global analysis of cinema advertisements in the *Düsseldorfer Nachrichten* between October 1918 and July 1921 allows us to put forward a few observations. First, there seems to have been little or no shortage of supply despite the dramatic political events occurring during these years. Programmes changed regularly and consisted mainly of popular and sensational titles, as well as films starring famous actresses and actors such as Henny Porten, Asta Nielsen, Harry Piel or Fern Andra. Series and serials played an important role in the programmes: the Residenz, which had close links to the Gaumont chain, screened almost all of Feuillade's serials: *Les Vampires* (France 1915/16), *Judex* (France 1917), *La Nouvelle Mission de Judex* (France 1918), *Tih-Minh* (France 1918), *Barrabas* (France 1919). Among the many more or less unknown and probably often non-preserved films, from time to time a title appears that is still well-known today, such as Joe May's serial *Die Herrin der Welt* (*The Mistress of the World*, Germany 1919) the four-part *Die Spinnen* (*The Spiders*) by Fritz Lang (Germany 1919/20), Robert Wiene's *Das Cabinet des Doktor Caligari* (*The Cabinet of Dr. Caligari*, Germany 1919) or Ernst Lubitsch's *Die Austernprinzessin* (*The Oyster Princess*, Germany 1919), *Madame Dubarry* (Germany 1919) and *Sumurun* (Germany 1920).

On the basis of the advertising it is difficult to determine the production countries of the films screened in Düsseldorf. It is, however, safe to assume that the majority were of German origin, and recent productions, although some older French productions, including *Les Vampires* and *Judex*, were released in Germany after the war. Similarly, D. W. Griffth's *Intolerance* (USA 1916) had its German release only after the war and was shown in Düsseldorf at the Residenz in December 1919.[48] Giovanni Pastrone's *Cabiria* (Italy 1914) was advertised by the Residenz in March 1921, the day after the occupation of the city. The advert highlighted the participation of Gabriele D'Annunzio and the enormous success of the film all around the world.[49]

All in all, it seems, the supply of films in Düsseldorf was little affected by the occupation.[50] Programmes changed regularly and offered a range of different titles. German films apparently circulated easily, with the exception, of course, of films banned by the Inter-Allied High Commission. Even here exceptions were possible. In a series of educational screenings for school children, Fritz Genandt showed on 25 November 1922 *Der Rhein in Vergangenheit und Zukunft* (*The Rhine in the Past and the Future*), one of the films that the Inter-Allied Commission had banned.[51] It is difficult to say whether such a special show escaped the attention of the military authorities or whether they considered an audience of school children inoffensive. The fact, however, that Genandt could get a print of the film into the occupied zone indicates that control took place at the level of programming rather than through the circulation of prints.

Conclusion

Looking at the situation in Düsseldorf during the years of occupation between 1919 and 1925, it is evident that the experience of cinemagoers was affected by the presence of foreign troops. Even if film availability was only marginally restricted by censorship measures taken by the Inter-Allied High Commission, the presence of Allied soldiers in movie theatres as well as the screening of prints with bilingual intertitles may have been perceived as something of an intrusion of the political situation into the realm of public entertainment. Other measures taken by the military authorities, such as the requisition of cinemas, bars and meeting rooms for their own moving picture shows provided different contexts for the consumption of films that also were attended by at least some of the local cinemagoers, who, in return, might have had their national loyalty questioned by their fellow citizens because of this.

On the basis of the documents available to us, we can reconstruct a general framework of actions, regulations and decisions by the Inter-Allied High Commission and a number of incidents that have left traces in the municipal archive of Düsseldorf or, in terms of cinema programmes, in the local press. Together, these provide insights into a largely neglected aspect of historical cinema cultures, namely the way in which political tensions – in this case the occupation of a region by foreign troops after the war – turn the everyday practice of cinema and the showing of individual films into contested objects. Both foreign and local authorities tried to keep some measure of control on the medium, but even where their goals might overlap in an abstract sense, their immediate national, political or commercial interests were often opposed.

Notes

1 See for instance the chapter dedicated to Weimar Cinema in W. Jacobsen *et al.* (ed.) *Geschichte des deutschen Films*, Stuttgart/Weimar: Metzler Verlag, 1993. Siegfried Kracauer's classic study of German cinema during the Weimar Republic, *From Caligari to Hitler* (Princeton: Princeton University Press, 1947) does not address this issue.

2 The authors would like to express their gratitude towards the staff of the Stadtarchiv Düsseldorf, who have been extremely helpful throughout.

3 *Agreement between the United States of America, Belgium, the British Empire, and France of the one part and Germany of the other part* [concerning the military occupation of the territories of the Rhine]. Stadtarchiv Düsseldorf (hereafter SAD) 0123.91.0000.

4 Ibid.

5 See for instance *Das Rheinlandabkommen sowie die Verordnungen der Hohen Kommission in Coblenz. Dreisprachige Textausgabe. Erläutert von H. Vogels und Dr. W. Vogels*, Bonn: A. Marcus & E. Webers Verlag, 1920. SAD 0123.92/4. All quotes from this source follow the English version given in this trilingual edition.

6 When the occupation ended in 1925, the German government had local and regional authorities in the occupied regions compile dossiers documenting what had happened during that period. The Stadtarchiv Düsseldorf holds a large number of files concerning the Rhine/Ruhr area. In addition, the government also edited material of that kind. See *Die politischen Ordonnanzen der Interalliierten Rheinlandkommission und ihre Anwendung in den Jahren 1920–1924. (Dokumente zur Besetzung der Rheinlande. Herausgegeben vom Reichsministerium für die besetzten Gebiete: Heft 1*, Berlin: Carl Heymanns Verlag, 1925). SAD 0123.94/5.

7 *Das Rheinlandabkommen*, p. 32.

8 *Benrather Tageblatt*, 20.6.1919 and 21.6.1919. Unless stated otherwise, all translations from German sources are ours.

9 *Die Stadt Essen im Ruhrkampf*, 11.1.1923–31.7.1925 (Essen 1927), p. 25. SAD 0123.108.0000.

10 *Denkschrift zum Ruhrkampf des Amts Datteln* (1926), p. 46. SAD 0123.103.000.

11 See the documentation and censorship lists about the occupation of the Rhine territories compiled by Herbert Birett on his website http://www.kinematographie.de/HCITR.HTM. After the Second World War, measures taken by the Allies concerning cinema were a lot stricter with regard to censorship, and in addition there was also a more active policy of influencing production and programming. For the French zone (which was further to the south and did not include Düsseldorf, which was part of the British zone), see Laurence Thaisy, *La Politique cinématographique de la France en Allemagne occupée (1945–1949)*, Villeneuve d'Ascq: Presses Universitaires du Septentrion, 2006.

12 *Das Rheinlandabkommen*, p. 92.

13 Ibid. See the English text of the High Commission's film-related ordinances on http://www.kinematographie.de/VOGB.HTM.

14 Ibid.

15 *Die politischen Ordonnanzen*, p. 14. The decision was published in *Amtsblatt der Reichskommission für die besetzten Gebiete*, 3, 17.1.1923, p. 14. The exact title of the film is *Der Rhein in Vergangenheit und Zukunft* (1922), an Ufa-Kulturfilm directed by Dr. Zürn, written by Felix Lampe. See H.-M. Bock and M. Töteberg (eds) *Das Ufa-Buch*, Frankfurt a. M.: Zweitausendeins, 1992, p. 68.

16 For these titles see http://www.kinematographie.de/FILMABC.HTM#F025.

17 *Amtsblatt des Reichskommissars für die besetzten Rheinischen Gebiete*, 13, 2. 4. 1921, p. 41. SAD 0123.93/7. See also http://www.kinematographie.de/AMTSBL.HTM#A128. For the credits of the film and a short summary see Bock and Töteberg, p. 86.

18 See http://www.kinematographie.de/LISTE.HTM. This film should not be confused with the American production *The Last of the Mohicans* directed by Maurice Tourneur, also produced in 1920.

19 *Amtsblatt der Reichskommission für die besetzten Gebiete*, 11, 1925, p. 25. The ban is announced in *Amtsblatt der Reichskommission*, 7, 1925, p. 1. See also http://www.kinematographie.de/FILMABC.HTM#F025.

20 *Denkschrift über den Ruhrkampf im Bereich der neuen Stadt Castrop-Rauxel*, 25.7. 1926, SAD 0123.102.0000.

21 *Benrather Tageblatt* (12.12.1921). Alternative titles for this film are *Der Fürstenmord, durch den Millionen starben* and *Die Schuld am Weltkrieg.*

22 See Ordinance 74, http://www.kinematographie.de/VOGB.HTM.

23 See http://www.kinematographie.de/HCITR1.HTM.

24 See the document dated May 21, 1921, http://www.kinematographie.de/HCITR3.HTM.

25 See Ordinance 123, http://www.kinematographie.de/VOGB.HTM.

26 We have not been able to find any concrete evidence of exhibitors programming foreign films in order to sidestep German regulations, so the concerns expressed by German authorities may have been hypothetical.

27 For general information about Düsseldorf during the period under investigation here see S. Lipski, *Dokumentation zur Geschichte der Stadt Düsseldorf. Band 2: Düsseldorf während der Revolution 1918–1919 (November 1918 bis März 1919). Quellensammlung*, Düsseldorf: Pädagogisches Institut der Landeshauptstadt Düsseldorf, 1985, and B. Brücher *et al.*, *Dokumentation zur Geschichte der Stadt Düsseldorf. Band 6: Düsseldorf während der Weimarer Republik 1919–1933. Quellensammlung*, Düsseldorf: Pädagogisches Institut der Landeshauptstadt Düsseldorf, 1985.

28 See *Material zu einer Denkschrift über die Besetzung Düsseldorfs (rechtsrheinisch) durch die Franzosen*. SAD 0123.123.0000. Information about the situation in Oberkassel mainly comes from O. Starck, *Die belgische Besetzung Oberkassels 1918–1926*, MA thesis, Heinrich-Heine-Universität, September 1995.

29 See *Entwurf zu einer Denkschrift über die Besetzung des rechtsrheinischen Stadtteils Düsseldorf durch die Franzosen*. SAD 0123.124.000.

30 See SAD III 7578 Militärkinos.

31 See *Entwurf zu einer Denkschrift*, op. cit.

32 Ibid.

33 See SAD III 7578 Militärkinos. There is in fact also a 1918/19 film with the title *Der rote Reiter* (dir. Fred Stranz), but given the circumstances the complaint quite probably refers to the 1923 production directed by Franz W. Koebner.

34 See ibid. Genandt may have been under particular suspicion from the German authorities as the French firm of Léon Gaumont owned a part of his company's shares.

35 In a letter to the Lord Mayor dated 29 July 1921, Commander Morin asked for a list of all the films shown that day, including a detailed listing of the items covered by the newsreel *Messter-Woche*. There are, however, no indications that there were any consequences to this action. See ibid.

36 See *Entwurf zu einer Denkschrift*, op. cit.

37 For this and the following see SAD X 261 Militär-Kino (1919). For a detailed account see S. Lenk, *Vom Tanzsaal zum Filmtheater. Eine Kinogeschichte Düsseldorfs*, Düsseldorf: Droste Verlag, 2009, pp. 55–56.

38 See ibid.

39 See SAD, XI 175 Lichtspiele (1920) and Lenk, *Vom Tanzsaal zum Filmtheater*, pp. 304–5.

40 See SAD X 261 Militär-Kino (1919) and Lenk, *Vom Tanzsaal zum Filmtheater*, pp. 59–61.

41 Ibid.

42 Ibid.

43 In addition, the foreign soldiers themselves may not have been too keen on mingling with the local population.

44 We would like to thank Eleni Giannakoudi for her invaluable help with regard to information about the films shown in cinemas in Düsseldorf during the post-war years.

45 According to the advertisements that appeared in the *Düsseldorfer Nachrichten*.

46 See advertisements in *Düsseldorfer Nachrichten*, 15.11.1918, 22.11.1918.

47 Ibid., 6.12.1918.

48 Ibid., 15.12.1919.

49 Ibid., 9.3.1921. *Cabiria* had passed the Berlin censorship in 1914 (no admittance for children, though). See H. Birett, *Verzeichnis in Deutschland gelaufener Filme. Entscheidungen der Filmzensur 1911–1920*, München: K. G. Saur, 1980, p. 187. We cannot say, however, to what extent the film was actually distributed and seen up to, and after the beginning of the war. The Residenz had shown an even older film, Enrico Guazzoni's *Quo Vadis?* (Italy 1913), at Easter 1919, with a full orchestra of 30 musicians accompanying the screening. So the question is whether *Cabiria* simply had a long shelf life and was successfully re-released occasionally, or whether this relatively old film was used to fill a gap in the supply of a distributor, caused by the political situation.

50 This, at least, is the impression one can get from the advertisements. How easy, or difficult, it was for the exhibitors to provide this offer of films to their clients is a different issue, especially since the Rhine–Ruhr area from time to time was cut off from the rest

of Germany, so that no goods could circulate. During the period of the occupation of the Ruhr in 1923, for instance, a report in *Der Kinematograph* (11.3.1923) states that the French and Belgian military confiscated all film material – 120 prints in one week – so that the supply was seriously compromised.

51 SAD III 2144 Einrichtung einer Bild- und Filmstelle sowie Lichtbildervorführungen (1922–26). This screening was an official one as it was organized in cooperation with the Bild- und Filmstelle, the office responsible for the municipal collections of films and images, the head of which, Hermann Boss, was also present.

4

'CHRIST IS COMING TO THE ELITE CINEMA'

Film exhibition in the Catholic South of the Netherlands, 1910s and 1920s

Thunnis Van Oort

During the 1920 spring fair in the Dutch town of Venlo, in the province of Limburg, the film *Christ* was screened at the Elite Bioscoop (or Elite Cinema).[1] Lambert Caubo, the son and designated successor of the cinema owner Jos Caubo, had arranged a spectacular publicity stunt for the show. He had hired a small airplane to disseminate pamphlets over the busy fair grounds, announcing in large letters: '*Christ* is coming', followed in tiny print by 'to the Elite Cinema'. Providence must have punished the hubris of the young Caubo, distributing flyers like manna from heaven, because the plane crashed into the backyard of a butcher, killing him and two others.[2]

This unfortunate incident could be taken as a metaphor for the deteriorating relationship between the cinema industry and Catholic authorities in the Dutch south in the early 1920s. Less than a year after the accident, the so-called Venlo Cinema War broke out, during which the four cinemas in the city closed for over a year rather than pay increased local taxes. This struggle revolved around the relationship between Catholic and non-Catholic cinema exhibition and distribution. Although local in origin, this conflict had a profound impact on the viability of Catholic cinema exhibition in the Netherlands. After this Cinema War, a daring publicity stunt such as that by Lambert Caubo, openly and ironically connecting religion to modern emblems such as aviation and the cinema, would become a lot less likely.

The province of Limburg was a relatively recent addition to the Dutch kingdom, annexed in the 1830s. Its almost homogeneously Catholic population did not have a strong sense of belonging to the predominantly Protestant Dutch nation. During the early decades of the twentieth century, a sudden boom in the coal mining industry in the southeast of the province accelerated Limburg's modernization. Concerned elites sought to control the socio-economic transformation of the region by a 'rechristianization' of the population, as part of a wider Catholic aspiration to gain more political power and influence within the Dutch state.

De Film

In de Elite-Bioscope, Venlo

FIGURE 4.1 The flyer distributed on the day the plane crashed performing a publicity stunt for Jos Caubo's Elite Cinema in Venlo, 1920

Cinema attendance in the Netherlands was among the lowest in Europe.[3] The metropolitan west and the Catholic south counted more cinemas and cinemagoers than most Protestant areas. Although reliable statistics are lacking for the 1910s and 1920s, Limburg appears to have been a relatively important market by Dutch standards. Nevertheless, the region had fewer than 20 cinemas in 1918, situated in only a handful of towns and cities with populations of between 10,000 and 60,000. Venlo, with approximately 20,000 inhabitants, had four cinemas in 1918.

Limburg provides a suitable focus for a study of cinema on a regional scale, particularly since the integration of cinema into the fabric of Limburg society took place in a dynamic interplay between local, regional and national levels. For instance, in 1918 a cinema owner from Maastricht (Limburg's capital) cried out in the national trade press for help against oppressive local government measures. This distress call prompted Dutch cinema exhibitors to form a national alliance to protect their interests. The Venlo Cinema War was a crucial event in the further development of the relationship between Catholics and the cinema industry during the interwar years. Although conflicts between cinema owners and local governments occurred elsewhere, the struggles in the Catholic south were longer, more heated and more public than in other parts of the Netherlands. The commercial exhibitors' victory in the Venlo Cinema War was commemorated in the Dutch trade press for years, serving as a trophy in the fight against government regulations, taxes and censorship.

The Venlo War and later confrontations between the interests of the cinema exhibition and distribution industry and Catholic government policies reveal the fault-lines in the Dutch cultural and political landscape. As in neighbouring Belgium, the Catholics were the most organized ideological group in their response to the alleged dangers of commercial popular culture. Although Belgian exhibitors loathed the level of Catholic interference and competition in their market, they did not engage in the open and prolonged conflicts that took place in the Dutch south.[4] In contrast to Belgium, institutionalized Catholic film exhibition in the Netherlands failed, partly because of the successful resistance of the professional organization, the

Netherlands Cinema Alliance (*Nederlandsche Bioscoopbond*), which had been established in 1918.

After the Venlo Cinema War, no Limburg cinemas were explicitly advertised as being Catholic. A scrutiny of local exhibition practices reveals that cinema entrepreneurs used a range of strategies, from silent negotiation to open conflict, in conducting their business and integrating cinema into a society that was more than 98 per cent Catholic. While some sought to negotiate a position in the local community for cinema by dressing it in respectable hegemonic notions of art or religion, others openly propagated the spectacular, emotional, and exciting aspects that made it attractive to many, but also threatening to others. In this chapter I will elaborate on the struggles between Catholics and the Dutch cinema exhibition and distribution industry, tracing how the conflicts over local and regional government control over the cinema in the early 1920s led to the establishment of a regional Catholic censorship organization that in turn facilitated the passage of national censorship legislation.

Cultural intermediaries

Cultural historians have adopted the concept of the cultural intermediary or broker from anthropology. It has been used successfully in early modern media history, for instance by Carlo Ginzburg and Robert Darnton in examining the influence of the printing press, but less frequently in film studies.[5] In brief, cultural brokers negotiate the appropriation of meaning across cultural boundaries. The French historian

FIGURE 4.2 An unidentified group of people posing in front of Caubo's Elite Cinema in Venlo. Judging from the visible film poster (announcing *Detective Swift*), the photo can be dated between 1914 and 1919. The group might well have rented Caubo's auditorium for a festivity

Michelle Vovelle introduced the term *intermédiaires culturels* in an attempt to make the borders between 'high' and 'low' culture more dynamic and permeable, by focusing on those people (or even objects) positioned between those spheres. The Catholic priest, responsible for transmitting and guarding values and rules from high up in the (cultural and ecclesiastical) hierarchy, would be a classic cultural intermediary. From time to time those rules and values would clash with local norms, and the priest would have to negotiate between the local and the official way. In adapting too readily to the local norms he might displease his superiors, but by clinging too strictly to the official rule he could alienate his flock.[6]

The concept of cultural intermediary allows us to focus on the role of individual actors (both persons and organizations) in integrating cinema culture into a social and cultural context, framing the way that entrepreneurs and policymakers negotiated the specific characteristics of cinema cultures in Limburg. Cinema owners were the most prominent intermediaries, acting as gatekeepers by deciding in what circumstances and in what manner and dosage his audiences were exposed to an international film culture, since in Limburg cinema was almost completely dependent on foreign input. The exhibitor chose marketing strategies and constructed the social image of his theatre. He balanced the demands of his audiences against the interests of his suppliers (the distributors) and the pressures of government and church authorities. The entrepreneur was also a final, critical influence on the way audiences experienced and interpreted the films they watched.

While some exhibitors remained close to the practices and tastes of their local culture, others allowed more room for the new and unfamiliar values and experiences brought by the cinema. Some exhibitors gave little sign of caring about the middle-class respectability of their theatre, concentrating their marketing strategies on excitement, emotion, the spectacular, the sensational, the erotic and the exotic. While these attractions were equally essential for their competitors, the distinction between them lay in the manner of their promotion. The open, unambiguous promotion of cinema's sensory attractions presented exhibitors with the serious risk of conflict with local authorities. Many exhibitors instead stressed the ideologically more neutral virtues of efficiency and convenience associated with urban modernity, and mixed this with an atmosphere of small-town cosiness and familiarity.

However they promoted their enterprises, all exhibitors experienced significant difficulty in legitimizing their cinemas by associating them with Catholicism. During the 1910s several Catholic cinemas did exist, but they either closed down or transformed into regular commercial cinemas. Cooperation beyond the local level did not occur, and the distance between cinema culture and Catholicism just seemed too wide for any direct and open alliance to be negotiated in the Netherlands. Because of the failure to associate cinema directly with Catholicism, some entrepreneurs sought other forms of social and cultural respectability, such as linking the cinema to theatre and other forms of high culture. Although this strategy was at times supported by international film production companies seeking a comparable legitimization for the film industry in general, it had only limited success.[7] For example, the first purpose-built cinema in Venlo, the Pollak Theatre (1925), was initially promoted as a theatre for respectable art. Although the majority of its income came from film screenings, they were given much less public attention; after a few years the theatre was taken over by a competitor with far fewer artistic aspirations.

Regional and national power struggles were largely fought on the local battlegrounds of provincial towns and cities, far away from the metropolitan areas. One of the central issues in recent debates on the history of cinemagoing, film exhibition and distribution, is how to connect micro-level to macro-level research. Examining Catholicism's attitude to cinema offers an interesting perspective on this question, since Catholic communities were simultaneously organized on local, regional, national and international levels. Although this chapter does not discuss the international dimensions of the changing relationship between Catholics and cinema, these were strongly influenced by the global restructuring of the film industry, especially the rising American dominance over the market after the First World War.

One way to address the problem of how to interconnect different scales of research is to locate and study the boundaries and connections between those different levels, focusing on the intermediary entrepreneurs and organizations involved in processes of negotiation between levels. Which entrepreneurs succeed in connecting their local business to regional or national organizations and networks? Were entrepreneurs who kept operating mainly in their local surroundings less successful? How did the Catholic censorship organization mediate in conflicts between national and local interests? Which incentives came from the 'top' and which came from the 'bottom'? Analysing those dynamics of scale can help in designing a framework for linking the micro and macro levels and studying the negotiations over the reception of international film culture in various communities. I would argue that the concept of the intermediary can facilitate transnational comparisons of how national differences in the organization of the exhibition and distribution industry and in legal and political arrangements regarding the cinema (such as censorship) brought about variations in the entrepreneurial styles and strategies in the repertoires of local exhibitors and other intermediaries involved in the 'acculturation' of cinema.

Catholic cinemas: private versus public

Catholic cinemas never really thrived, as far as can be judged from the sparse information that has been recovered from archives, newspapers and business records. They were always troubled by the uneasy compromise between cinema's commercial potential and its moral danger. Catholic cinemas existed in several cities during the first decade of fixed location cinema in Limburg. The first such venue opened in 1911 in the mining town of Heerlen. Like other cinemas, Catholic venues appeared within the existing infrastructure of the entertainment industry, in this case organized by Catholic associations such as the Saint Pius Association in Venlo, or the Catholic *patronaten* (confraternities) in Maastricht and Heerlen, organizations that were founded in order to protect the working class youth from moral decline.

The fact that these Catholic cinemas were voluntary private associations triggered the first problems with the regular exhibitors, who ran businesses that were legally public institutions of entertainment. In the early twentieth century commercial leisure was for the most part structured as a loose-knit sociability network, consisting of *verenigingsleven* or voluntary associations. Early cinema exhibitors cooperated with local organizations, many of them monitored by Catholic clergy and often supported by Catholic authorities. These leisure entrepreneurs derived a significant income from renting out their auditoria for meetings, concerts, plays and other festivities organized by the local associations. As the popularity of cinema increased during and especially immediately after the First World War, however, daily film screenings pushed out less lucrative club activities. This caused tensions between the cinema industry and local social organizations. This rift was widened by the influx of products that were perceived as immoral, including the German *Aufklärungsfilme* (sensational so-called educational films about tabooed sexual topics) that were very popular during the hectic episode following the end of the war.[8]

Even more problematic was the fact that Catholic clubs started to use their privileges as voluntary associations in competition with regular, public cinemas. Moreover, these Catholic initiatives were often supported by local governments dominated by Catholic politicians and clergy. Some clubs saw an opportunity in the post-war cinema boom, and set up their own screenings. Although little is known about these Catholic cinemas' programmes or their audiences, they were sufficiently competitive for the regular exhibitors to protest vehemently against them.

With the rise of cinema as a permanent site of entertainment, local governments developed regulations on entertainment taxes, censorship and age restrictions. The first local laws in Limburg came into effect from 1913 onwards. In contrast to the regular public cinema owners, private associations were not subject to censorship measures, nor did they have to pay entertainment taxes, reducing their costs by as much as 30 per cent of ticket sales. Finally, they were not hindered by the age restrictions that were introduced in many cities and towns. While regular exhibitors found that an important segment of their audiences was being banned from their premises through these age restrictions, most Catholic cinemas appear to have specialized in young audiences. Officially, screenings in the Catholic cinemas were private, but in practice they appear to have functioned quite similarly to public cinemas, admitting non-members as well. The private/public distinction was used as a legal loophole by local governments sympathizing with Catholic cinemas.

Venlo Cinema War and censorship

This situation led to the 1921–22 Venlo *bioscoopoorlog* or Cinema War. When the city council raised local entertainment taxes at regular cinemas to 30 per cent, the two regular Venlo cinemas, with the support of the Netherlands Cinema Alliance, decided to close in March 1921. The now unemployed exhibitors were supported from a strike fund. The Alliance's representatives negotiated with the city council, but to no avail. The conflict dragged on for months.

The Alliance had arranged a complete boycott of film supplies to the town's cinemas, including the two Catholic cinemas. During the summer of 1921, the Alliance significantly increased its power base through a merger with the distributors, forming an interest group in which exhibitors and distributors were united. The Alliance used the Venlo conflict as a demonstration of its power, internally to exhibitors and distributors and externally to local and national governments.[9] After an attempt to break the boycott by the Catholic Workers' Cinema failed, the city council gave in to the Alliance's demands, the tax was returned to its original level and the same regulation was applied to regular and Catholic cinemas alike.

Catholic cinemas had now lost their privileges as private institutions, and the Catholic Workers' Cinema ceased to exist. The Saint Pius Cinema remained in existence for some years as Limburg's only Catholic cinema, as other Catholic cinemas in the region closed, or were transformed into regular, commercial businesses without explicit reference or connection to their previous Catholic capacity. After the Venlo conflict, Catholics did not succeed in setting up an economically viable exhibition infrastructure. In the 1930s a modest circuit of Catholic film shows predominantly serving villages without fixed-location cinemas was created, but this never became a commercial threat to the regular film industry.

Following the failure of Catholic film exhibition, attention shifted to regulating the distribution of film in the Catholic south by setting up a censorship apparatus. Censorship existed on the local level: each city and town had its own regulations and committees. From 1915 Catholics had attempted to organize film censorship beyond the local level, with the publication of the periodical *Tooneel en Bioscoop* (*Theatre and Cinema*), edited and published in Rolduc, a prestigious Limburg institution for the education of priests. This publication consisted of lists of film and plays rejected or tolerated according to orthodox Catholic moral standards, and also listed the cuts that individual films required to be acceptable. This journal was adopted by a growing number of Catholic city councils as a legal register determining which films were permitted for screening in the local cinemas.

The regional collaboration that this periodical inspired was further stimulated by a national government increasingly dominated by Catholics in confessional coalitions with Protestants. In 1918 the Limburger Charles Ruijs de Beerenbrouck became the first Catholic Prime Minister of the Netherlands, and took the first steps towards national cinema legislation by installing a 'state committee on the subject of measures combating the moral and social danger connected to cinema exhibition'.[10] In 1920 this committee recommended the passage of a national cinema act. The act was not, however, drafted until 1923, and it was then rejected by parliament. It was another five years before any such legislation was enacted.

The passage of national legislation took too long for the mayors of several important Catholic cities in the south, who founded the Southern Film Censorship Association (hereafter called Censorship Association) in 1923.[11] Their initiative was supported by the Catholic Prime Minister, and thus from the outset linked to national political objectives. The association was in fact a club run by mayors, coordinating the reviewing, censoring or even banning of films in the Catholic south. City councils could vote for their city to become a member of the association, thereby subjecting the cinemas in their community to the verdicts of the censor. Beginning with 18 members, the association included 37 cities and towns in 1939, covering the great majority of communities in the Catholic south of the Netherlands in which cinema exhibition took place.

The responsibility for judging the film supply for hundreds of thousands of cinemagoers was carried by a single censor, Bernard De Wolf, a Catholic journalist and former Amsterdam city council member, who reviewed all the films either in his Amsterdam office or in the distributor's showrooms. Large business interests were at stake in his decisions, and after several years it emerged that De Wolf was not as objective as he might have been, since he was secretly paid by many distributors for creating approved versions of films that otherwise would have been banned from the considerable Catholic market. In 1926, for instance, great commotion was caused by De Wolf's approval of Sergei Eisenstein's *Battleship Potemkin*. The Censorship Association endured harsh criticism from its members for allowing a film that was seen by many authorities as a communist provocation made by a sworn enemy of Catholicism. The crisis was solved with difficulty, by taking special measures such as adding extra intertitles warning the audience about the pernicious effects of a socialist revolution and repeatedly praising God and Queen to ward off the red danger.

The Netherlands Cinema Alliance was, not surprisingly, hostile to the foundation of this local censorship association. Several conflicts, often on a regional scale, broke out between the two organizations. When the Censorship Association began its operations, the Cinema Alliance closed all cinemas in the cities under the authority of the association, using its boycott in what soon became a power struggle between the industry and the government over the control of the cinema business. After only one week of cinema closures, a relatively quick compromise was reached, in part because the Cinema Alliance recognized that there were some commercial benefits to regional censorship. Local censorship committees were capricious in their decisions, disadvantaging exhibitors and distributors when a title was banned at short notice before the scheduled public screening. Centrally organized censorship made business much more predictable and therefore profitable.

A second conflict lasted considerably longer, however. In 1928, national legislation was finally adopted, regulating the relationship between government and cinema exhibition and introducing a national board of review. The Cinema Alliance considered that a separate Catholic censorship body was now redundant as well as bothersome, since each title now needed two reviews, occupying costly time during which the film could not be rented out. To Catholic authorities in the south, however, the national board of review could not guarantee Catholic moral standards. Prime Minister Ruijs de Beerenbrouck had made sure that the 1928 Cinema Act provided a loophole for local re-reviewing. Although the centralized Censorship Association was in practice not operating on a local level, it could nevertheless legally remain in operation. The Cinema Alliance vehemently opposed this arrangement. In 1928 the corruption of the Association's censor De Wolf was exposed in order to incriminate the Association as a whole, but this only resulted in a further deterioration of the relationship between the two parties. When the Cinema Alliance maintained its refusal to accept Catholic censorship, Catholic city authorities started closing cinemas in the summer of 1929. In retaliation, the Cinema Alliance started boycotts

DE HEER B. TH. DE WOLF,
redacteur van »Tooneel en
Bioscoop.«

FIGURE 4.3 Film censor Bernard de Wolf

in the south. For five months, about 25 cinemas in 12 southern cities closed down. On 24 October 1929, Prime Minister Ruijs de Beerenbrouck intervened personally to end the conflict. On this occasion, the Cinema Alliance was not victorious as it had been in the 1922 Venlo War. The upheaval in the industry surrounding the introduction of the sound film did not permit a prolonged and costly battle against Dutch authorities. The agreement of 1929 formed the basis of a more peaceful coexistence between the distribution and exhibition trade and the Catholic authorities during the following decades, and separate Catholic censorship remained in place until 1967.

It would be misleading to see the Catholic response to cinema as only anti-modernist and reactionary. Catholics were the most organized ideological group in the Netherlands, partly because of their use of modern media such as press and radio. Cinema proved to be more difficult to manage, however, in large part because the overwhelming majority of film production took place abroad, and because the Dutch cinema market was small and strongly centralized in the Netherlands Cinema Alliance. Unlike radio or the press, film screenings depended on publicly accessible but privately owned venues that were difficult to put under direct Catholic control.

FIGURE 4.4 *The hand of the censor.* Cartoon by Albert Hahn commenting on the Cinema Act in the left–wing periodical *De Notenkraker* (1925)

Exhibitors as intermediaries

The 1920s saw a gradual shift from local to supra-local intermediaries. During the 1910s the exhibitors, local politicians and clergy were the most prominent cultural brokers shaping cinema culture in their towns. With the growth and professionalization of the cinema industry, supra-local organizations such as the Netherlands Cinema Alliance and the Catholic Censorship Association increased their influence. Key cultural intermediaries were operating on regional and national scales, such as censor De Wolf, the politicians in the Censorship Association, and the businessmen in the Cinema Alliance. This shift was reflected in the way that local exhibitors were viewed. Many Catholic authorities regarded exhibitors as helpless victims forced to screen the depraved products supplied by national distributors and international producers, while distributors considered their local clients to be just as much as the victims of their paternalistic governments. With the increasing influence of these supra-local organizations, the space that local exhibitors had to manoeuvre as intermediaries grew narrower, but did not disappear altogether, especially if exhibitors succeeded in connecting to wider networks.

Willem Peters, who opened the second cinema in Venlo in 1912, provides an example of an exhibitor who operated as a local intermediary. Peters promoted his cinema as a venue of popular entertainment. Unlike some of his competitors, Peters did not attempt to package his theatre as a temple for art. He resented the efforts by Catholics to control the cinema business, and he played a crucial role in the Venlo Cinema War in opposition to the local government and the Catholic cinemas. In this fight he came into contact with the Netherlands Cinema Alliance, and later became a member of its board. Through his contacts in the Alliance, he was able to finance the take-over of a rival cinema in Venlo in 1930. By that time he was the most successful exhibitor in Venlo, and hardly bothered by competition. Peters operated aggressively, opposing local authorities in several conflicts and not attempting to camouflage but openly advertising the sensational nature of cinema culture. This approach was apparently successful, for Peters became one of the most successful cinema exhibitors in the region, in part as a result of the support he received from the national Alliance.

Peters' approach was not the only route to success, however. Most exhibitors preferred peaceful compromise. In the booming mining town of Heerlen, local intermediaries hardly operated outside their local network and were nevertheless successful. Anton Weijerhorst managed to monopolize cinema exhibition in Heerlen for the greater part of the interwar years, without coming into conflict with the local authorities. The local government even cooperated with the exhibitor in the building of a municipal theatre intended to uplift the inhabitants through legitimate theatre and highbrow music performances. The theatre could only finance itself by operating a cinema during the weekends. On 24 October 1930, the town celebrated the theatre's opening night with a film that could hardly be considered a serious contribution to high art: the MGM production *The Flying Fleet* (1929) starring Latin lover Ramon Novarro. The town of Heerlen was a somewhat exceptional case, however. The municipal government was critical of the polemic stance of the

Het symbool der kracht, die uitgaat van het
Bio-vacantie-oord. De heer Peters was zoo
welwillend op dezen schoonen dag als symbool
te poseeren.

FIGURE 4.5 Portrait of Venlo exhibitor Willem Peters used to illustrate an article in the
trade press journal *Nieuw Weekblad voor Cinematografie* (3 April 1931).
Peters poses as a symbol of 'strength' for a charity project initiated by the
Netherlands Cinema Alliance. The portrait illustrates Peters' close involvement in the Alliance.

Censorship Association, and eventually cancelled its membership, in favour of a less stringent local cinema policy. Weijerhorst's entrepreneurial strategy to steer away from conflict with the authorities was part of a mutual concern to regulate the market, just as much as Willem Peters' confrontational style was a response to a much more bellicose local government.

In the Netherlands, the cinema infrastructure developed outside the realm of Catholic organizations. The Censorship Association's control exerted some influence over distribution, preventing a minority of the most 'immoral' films entering the region. But in the end this was largely a symbolic gesture, suggesting to the Catholic rank and file that the cinema was under control. One could even argue that the Catholic censorship apparatus unintentionally helped to endorse a commercial cinema culture, by allowing the majority of films distributed in the Netherlands to be screened.

The fact that Catholicism and cinema could not be reconciled into some hybrid form of commercially viable Catholic exhibition points to the limits of cultural mediation after the conflicts in the early 1920s. Apparently the gap between official Catholicism and commercial cinema was too wide in the Netherlands for any cultural intermediary to cross successfully, at least in the open. In Limburg's capital, Maastricht, two prominent Catholic businessmen opened the Mabi cinema in 1930. One of the owners, Johannes Teunissen, was the accountant of the episcopal administration. Their cinema did not carry any explicit Catholic signs and lacked any official affiliation to Catholicism, but the exhibitors were extremely careful in their programming strategy not to screen offensive films. They cut undesirable scenes when this was deemed necessary, without the consent of the distributor, even after national and Catholic censorship had been applied. On days when a religious procession would pass the cinema, a large statue of an angel was placed on the front balcony.[12] A decade after the unfortunate plane accident in Venlo, distributing flyers announcing the coming of *Christ* to the Elite Bioscoop, this was perhaps the closest Limburg would get to advertising Catholicism in the cinema.

Notes

1 This 1920 German film by Ludwig Beck was originally titled *INRI – Die Katastrophe eines Volkes* (*INRI – The Catastrophe of a People*). See S. Lenk, 'Rekonstruiert: *INRI – Die Katastrophe eines Volkes*', *Filmblatt* 3, 1998, 26. The Dutch title was simply: *Christus*.
2 For detailed references to (Dutch) primary sources, see my dissertation: T. van Oort, *Film en het moderne leven in Limburg: Het bioscoopwezen tussen commercie en katholieke cultuurpolitiek (1909–1929)*, Hilversum: Verloren, 2007.
3 K. Dibbets, 'Het taboe van de Nederlandse filmcultuur. Neutraal in een verzuild land', *Tijdschrift voor Mediageschiedenis* 9, 2006, 46–64.
4 D. Biltereyst, 'De disciplinering van een medium. Filmvertoningen tijdens het interbellum', in D. Biltereyst and Ph. Meers (eds) *De verlichte stad: Een geschiedenis van bioscopen, filmvertoningen en filmcultuur in Vlaanderen*, Leuven: LannooCampus, 2007, pp. 45–63, 58.
5 C. Ginzburg, *The Cheese and the Worms: The Cosmos of a Sixteenth-Century Miller*, New York: Penguin, 1982. R. Darnton, *The Great Cat Massacre and Other Episodes in French Cultural History*, New York: Basic Books, 1984. See also M. Vovelle, 'Les intermédiaires culturels: une problématique', *Colloque 'Les intermédiaires culturels', organisé par le Centre*

Méridional d'histoire sociale des mentalités et des cultures, Aix-la-Baume, June 1978. R. Maltby, 'On the Prospect of Writing Cinema History from Below', *Tijdschrift voor Mediageschiedenis* 9, 2006, 74–96.

6 The Dutch historical ethnologist Gerard Rooijakkers analysed this and other dilemmas of the cultural intermediary in his dissertation on early modern religious life in the south of the Netherlands. G. Rooijakkers, *Rituele repertoires. Volkscultuur in oostelijk Noord-Brabant 1559–1853*, Nijmegen: SUN 1994.

7 W. Uricchio and R.E. Pearson, *Reframing Culture. The Case of the Vitagraph Quality Films*, Princeton: Princeton University Press, 1993.

8 See for instance M. Hagener and J. Hans, 'Von Wilhelm zu Weimar. Der Aufklärungs- und Sittenfilm zwischen Zensur und Markt', in M. Hagener (ed.) *Geschlecht in Fesseln. Sexualität zwischen Aufklärung und Ausbeutung im Weimarer Kino, 1918 – 1933*, München: Text + Kritik 1999.

9 In the town of Hilversum a similar, although smaller conflict was fought simultaneously between local authorities and exhibitors backed by the Alliance.

10 R.B. Ledeboer, *Verslag der staatscommissie in zake maatregelen ter bestrijding van het zedelijk en maatschappelijk gevaar aan bioscoopvoorstellingen verbonden*, The Hague, 1920. J. van der Burg and J.H.J. van den Heuvel, *Film en overheidsbeleid. Van censuur naar zelfregulering*, Den Haag: SDU 1991.

11 The official Dutch name of the association was: *Vereeniging van Noord-Brabantsche en Limburgsche gemeenten ter gemeenschappelijke filmkeuring*.

12 F. Jansen, *Bewegend beeld. Bioscopen in Maastricht vanaf 1897*, Maastricht: Stichting historische reeks Maastricht, 2004, pp. 89–90.

5

IMAGINING MODERN HUNGARY THROUGH FILM

Debates on national identity, modernity and cinema in early twentieth-century Hungary

Anna Manchin

Film scholars and cultural theorists have explored the relationship between the emergence of cinema and the experience of modernity in various early twentieth-century settings.[1] Part of this project has been the investigation of how cinemagoing contributed to the creation of new social structures and expectations among audiences. Several recent studies have begun to complicate the meaning of cinema's modernizing potential, emphasizing the ways in which cinema interacted with and was integrated into pre-existing structures and cultures, not all of which can easily be characterized as traditional.[2]

My study adds to this discussion by emphasizing the political dimensions of the debate on cinema's modernizing potential. The idea that films could create a new society or – seemingly overnight – transform traditional culture was a fantasy entertained by contemporary detractors and cinema enthusiasts alike. Critics of all political persuasions debated the dangers and potentials of cinematic modernity, insisting on the power of the moving image to shape audiences and the effects that cinemagoing would have on traditional society and culture. All of these discussions of cinema and audiences took place in the wider context of political debates on modernity. The ways in which critics thought about the effects of cinema were conditioned by their attitudes towards 'modernity' and 'tradition' in the specific political context. Conversely, arguments concerning the changes brought on by cinema helped to establish the meaning of commercial modernity for Hungarian progress and the place of traditional culture in national identity.

Cinemagoing in early twentieth-century Hungary

Considering early twentieth-century Hungary's image as a backward, traditional and rural part of the Austro-Hungarian monarchy, the seemingly quick and easy advance of cinema here is at first surprising. Films reached Hungary in the spring of 1896, shortly after they appeared in Paris. The Projectograph, the first permanent movie theatre in Budapest, opened in 1906 and within a decade or so had hundreds of competitors.

In fact, the conditions in turn-of-the-century Budapest were almost ideal for the development of a culture of cinemagoing. Budapest was a rapidly growing cosmopolitan centre with a large immigrant population eager for entertainment; it was a centre of avant-garde culture and had an entrepreneurial middle class willing to venture into new areas of business including the mass press, popular sport and cinema. By the turn of the century, Budapest also had a well-established network of night-time entertainment. A city notorious for its cabarets, theatres, Orpheums, brothels and a large informal public sphere of coffee houses and newspapers, Budapest incorporated films into its pre-existing commercial structure. Before the appearance of permanent movie theatres, films were often exhibited at these entertainment establishments. Entrepreneurs with some experience in commercial entertainment opened Budapest's first cinemas, which spread rapidly across the city centre. As a December 1906 article in the daily *Vasárnapi Újság* put it, 'almost every week, a new cinema opens'.[3] By 1910, a film industry journal pronounced Budapest, a city famed for its coffee-house culture, 'no longer the city of coffee houses, but rather the city of cinemas'.[4]

At the peak of the cinema boom in 1914, Budapest had 108 movie theatres. Although this number declined in the following two decades (fluctuating between 77 and 87 throughout the 1920s and 1930s) the total seating capacity of Budapest theatres gradually increased from 29,962 in 1914 to over 40,000 by the late 1920s.[5] In 1927, for example, an average of 37,000 cinema tickets were sold every day in Budapest.[6] Hungary also developed its own film industry, which was based almost entirely on small and micro-investments until the Second World War. By the last years of the First World War, Hungary became one of Europe's largest film producers, exporting to countries that banned American films. The first generation of Hungarian filmmakers included Alexander Korda, Maria Corda, Paul Fejos, Ladislao Vajda and Michael Curtiz.

Budapest, the country's financial, cultural and administrative centre, was an anomaly in Hungary, which was a rapidly modernizing but still largely rural country with over half its population employed in agriculture. As elsewhere in Europe, films initially reached the countryside as part of travelling shows and were later shown in coffee houses or traditional theatres.[7] Even in 1928, only 10 per cent of Hungary's settlements (368 out of a total of 3,459) had movie theatres, and this percentage did not change drastically in the 1930s.[8] Although provincial cinemas accounted for much of the increase in the number of theatres across Hungary in the interwar decades (from 367 in 1921 to 464 in 1939), this increase was relatively small.[9] The lack of a solid audience base made it difficult to sustain a programme and to justify investment, and as an additional setback, a 1928 government decree explicitly prohibited towns under 1,000 inhabitants from acquiring movie theatre licences except by special permit from the Ministry of Interior Affairs.[10] The decree stated that small rural settlements were politically and culturally 'immature': authorities were concerned that cinemagoing could offer occasion for social and political unrest. The decree also exemplified the conservative belief that 'authentic' national culture in the rural countryside had to be protected from the cultural effects of urban mass culture.

Of Hungary's 367 movie theatres in 1921, 91 were in Budapest, 67 in the area around the capital, and 209 in the countryside. Although only about 11 per cent of the population lived in Budapest, the city had nearly 25 per cent of all movie theatres, while another 20 per cent were in the surrounding areas, including the suburbs.[11] By 1935, this ratio had shifted considerably: only 17 per cent of movie theatres were in Budapest and 75 per cent in the countryside; 60 per cent (310 movie theatres) were in small towns.[12] While most theatres in Budapest had screenings every day of the week, usually several times a day, rural theatres typically showed films once or twice a week with a few shows a day. Over half of Hungary's population lived in a settlement without a permanent cinema, although fewer people lacked access to movie theatres entirely, as many travelled to nearby towns in search of urban entertainment forms, particularly on the weekends.

Despite the discrepancies between urban and rural cinemagoing, films became the most popular commercial entertainment all over early twentieth-century Hungary. Traditional theatre was much more expensive and largely unavailable outside cities. No other popular entertainment form came close to cinema in terms of popularity and accessibility: cinema gradually replaced all competing attractions such as circuses and side-shows. Radios, while widespread in cities by the late 1920s, remained a middle-class luxury in the countryside until the start of the Second World War: only 1 in 30 households had one. According to 1935 statistics, Hungarians spent almost exactly as much on movie tickets that year as they did on all other printed materials (newspapers, magazines, books, calendars, etc.) combined.[13]

The meaning of moviegoing

What to make of this new culture of cinemagoing and how to interpret the advance of films? Several narratives were possible, and they were closely related to the competing definitions of modernity that emerged along the political spectrum in turn-of-the-century Hungary. There was general agreement that Hungary was backward and lagged behind in comparison to Western Europe, and critics concurred that this problem needed an urgent solution. Since the mid-nineteenth century, liberal reformers had pushed Hungary to follow a Western model of modernization, insisting that industrial capitalism, liberalism and social change offered the only route to progress.

A different view of Hungarian modernity emerged from the conservative side; gaining strength at the turn of the century and becoming increasingly powerful in the face of the crisis of liberalism. Worried that Hungary was losing its cultural autonomy, conservatives sought a modernity based on national uniqueness and traditional culture, under the leadership of Hungary's traditional elite, the landowning nobility. They wished to protect the traditional social order against the incursions of liberalism, democracy and commercial urban culture, all of which they regarded as inauthentic foreign developments detrimental to authentic national progress. The turn-of-the-century desire for a renewal based on a return to 'authentic' values was not unique to Hungary, but here the conflict between tradition and modernity assumed an ethnic and racial tone.

Hungary's industrial, financial, commercial and professional bourgeoisie was largely of Jewish origin. Although by the early twentieth century, the Christian and Jewish upper and upper middle classes were inextricably connected through business and family ties, capitalist, urban modernity was understood as being Jewish.[14] The dangers and threats that capitalist modernity posed to the continuity of Hungarian national traditions and culture were also seen as foreign and Jewish.[15] As Pal Ignotus, a liberal Jewish intellectual, satirically explained,

> It would be useful to collect how many things were deemed non-Hungarian in the past decade. Budapest is not Hungarian. The dialect of Pest is not Hungarian. The stock exchange is not Hungarian. Socialism is not Hungarian. Internationalism is not Hungarian. The organization of agricultural workers is not Hungarian. Capital is not Hungarian. Secession and symbolism are not Hungarian. It is not a Hungarian idea to exclude the religious institutions from public education, or to eliminate religion from the curriculum. Caricature is not Hungarian. Greater tolerance towards love is also not Hungarian. General Suffrage is not Hungarian. Materialism is not Hungarian and it is not Hungarian to suppose that people may change their institutions … rationally, according to their needs. But most of all, whoever is not satisfied with the existing situation is certainly not a Hungarian and such a person should have the good sense to leave the country with which he is unsatisfied.[16]

Before 1914, conservative nationalism failed to gain dominance in public debates and political life, but this changed after the traumatic events of the First World War. After the war, Hungary lost two thirds of its former territory and population, and was transformed from a member of a multinational empire to a small and fairly homogeneous nation state. After a bourgeois democratic revolution in 1918 and a brief communist dictatorship in 1919, Hungarian public life took a conservative turn and a new authoritarian government came to power. Led by Admiral Miklós Horthy, the regime defined itself through the rejection of all radical social change, liberalism and democracy, and sought to institutionalize a definition of authentic Hungarian national culture as rural, traditional and Christian. From the end of the war, Jewish Hungarians and the urban, commercial modernity they supposedly brought into being were imagined as the root of all of the nation's problems, and ironically, as the cause of its backwardness.[17]

Until the Great War, the debate on films and their audiences took place in the context of a hopeful view of national commercial modernization and democratization. Proponents welcomed films, arguing that they would propel Hungary towards Western, cosmopolitan modernity. Although the conservative political elite, resistant to social change, regarded movies as potential threats to the traditional social order and accordingly wished to restrict their influence, movies were not considered a serious concern on a national level before the First World War. Regulation of movie theatre permits and all aspects of screenings were left up to local authorities, and there was no official film censorship until 1920.

In the interwar period, a hopeful view of modernity was replaced by a profound distrust of cosmopolitan, commercial modernization. The problems of modernity in public life and politics came to be discussed as a 'Jewish problem': critics saw modernity in Hungary as 'foreign' and 'Jewish' and called for an authentic national or Christian version. This cultural and political context defined the parameters of the debate on the modernizing influence of films. The conservative government and radical nationalists sought to reshape the film industry and films themselves to better fit this new vision of the modern nation, at the same time arguing that films would help. Paradoxically, however, far from bringing a new kind of conservative national community and national cinema into existence, films remained the most important public expressions of a commercial, democratic alternative to the dominant conservative racist national modernity until the mid-1930s.

Metropolitan dreams in provincial Hungary: film exhibitors and producers on the cultural role of cinema

In the early twentieth century, it was avant-garde artists, liberal intellectuals and writers who were most enthusiastic about the democratic and modernizing potentials of cinema. Endre Ady, Georg Lukács, Lajos Biró, Béla Zsolnai, Dezsō Kosztolányi, Zsigmond Móritz and many of the writers for Hungary's most influential modernist journal *Nyugat* (*West*) felt that the social and cultural experience of cinema and cinema's universal language would help bring about a new cultural and social world order.

Although not all cinema enthusiasts celebrated such radical change, before the First World War, film industry professionals often discussed cinema's positive, civilizing influence on the development of Hungarian culture. As the following examples make clear, in the first two decades of the twentieth century, a liberal vision of modernization based on Western, cosmopolitan capitalism could still be explicitly discussed as national: urban modernity and national identity were not always understood to be in conflict. This reflected a general agreement among the early twentieth-century public that Hungary was lagging behind the West and had to catch up and modernize following a Western capitalist model; rural Hungary, in this view, was a priority and would have to change most rapidly.

In 1907, Elemér and Rezsō Suchan applied to the city council for a permit to establish an 'electric theatre' in Debrecen, a small town on the Great Plains of East Hungary. In their application, they argued that:

> All larger cities in civilized, cultured Europe try their best to entertain their residents in a morally noble fashion, distracting them from often more dangerous pastimes. This is why theatres developed in the old days, followed in most recent times by enterprises that visually entertain and educate audiences undifferentiated by gender or age. ... Morally and financially, these institutions are invaluable. This is the only place where such cheap and edifying entertainment can be offered to the population. This is why other similar electric theatres were brought into being in London, Paris, Berlin, Vienna and Budapest.[18]

There are several claims here worth elaborating upon. First, Suchan and Suchan posited 'larger cities in Europe' such as London, Paris, Berlin and Vienna as paragons of culture and civilization, and they imagined Budapest, and even possibly Debrecen, to be among these European metropolises. To them, culture and civilization appeared as universal; cultural progress had the same meaning and could take the same route in all European nations. Second, in contrast to conservative critics' view that cinema had a detrimental effect on the social order and public morals, they presented movie theatres as having a positive effect on morals, culture, and society, likening their own role to that of the classical theatre. Third, they celebrated the democratic expansion of the public to women and to younger and older audiences as a positive development and as a sign of progress. They imagined cinema as bringing to the nation modernity and cultural progress, defined as urban, commercial, and democratic.

A similar perspective was expressed by Mihály Meixner, a coffee house owner in Győr, a small town in the north-west. In July 1907, Meixner petitioned the city for permission to screen films outdoors, projected onto a wall across the street from his establishment, arguing that this would allow even individuals unable to afford drinks at his café to enjoy the films. 'My suggestion is not new; this practice is well-known in Budapest, Vienna, and in large foreign cities. It would serve to heighten the metropolitan character of Győr while at the same time enhancing its commerce.'[19] In Meixner's view, a 'metropolitan character' as well as an increase in 'commercial' culture were unquestionably desirable things for a city; he believed Győr ought to compete with and follow the examples of larger European cities. Meixner also believed that his individual business could provide cultural benefits to the city by further expanding the public reach of cinema's civilizing and educational potentials. The city council rejected his proposal, however, in part because of the alleged dangers it held for public order. 'Summer shows would draw in audiences from the outskirts', the council explained. 'Since it would be impossible to separate out immoral elements, the police would be incapable of maintaining order.'[20]

A 1906 article in Győr's daily paper, the *Győri Hirlap*, had lauded the effects of movie exhibition and coffee houses on cultural and urban progress in terms similar to Meixner's:

> Night life in Győr is increasingly gaining a really metropolitan character because of the movie theatres. [The movies] can be seen in three coffee houses, and all of them are packed with the type of audiences that until now did not seek out night entertainment at all. Spaces in which moving images are exhibited are frequented by whole families; this is proof of the moderate, intelligent, and edifying character of such entertainment. It would be more correct to say that the moving image exhibitions in our city have established cheap, family-friendly and edifying entertainment … [One coffee house owner who exhibits movies] has mobilized nearly 10,000 coffee house guests.[21]

Here again, urban progress and metropolitan culture appeared as desirable, as movies provided moral entertainment to an expanding public that included families and the poor. Cinema brought civilization by drawing people who had previously been excluded from the nation's cultural circuit into the public sphere.

At the opening ceremony of the new Uránia movie palace in Nagykaniza in October 1912, the owner, Nathan Armuth, a wholesale merchant from the town of Kanizsa, promoted the beneficial effects of movies:

> We can consider movies a veritable blessing, especially in smaller, more isolated rural areas, where there are few other cultural opportunities open to the general public. But now the movie theatre is here, and this means everything for the poor man's culture. For 60 *filler*, they can see the U.S. presidential elections, any Wagner opera with piano accompaniment, and even if this experience will not amount to real art appreciation, [audiences] will at least know of the great cultural accomplishments that take place in the world.[22]

To Armuth, distance from international high culture resulted in cultural poverty, backwardness and exclusion. Civilization was possible only through a common Western culture, which the new technological advances in film made accessible to the poorest segment of society. He considered this the highest purpose of entertainment and a boon to the nation.

The image of an ideal modern Hungary suggested by these proponents of cinema was one of an increasingly urban place where accessible mass culture offered a common experience for everyone and contributed to social and cultural progress. National modernity was to be achieved through commercial culture, urbanization and economic development.

Movie theatres, functioning as public institutions, participated in shaping an understanding of the modern Hungarian nation. If films were capable of changing audiences' notions of time and space, so too were cinemas, which often did so from a particular political perspective. For example, cinema owners displayed their patriotic sentiment by marking public holidays with changes in their exhibition schedules or programming. When the Austro-Hungarian royal couple were assassinated in 1914, most exhibitors altered their schedules with special programming – usually with a day of mourning; this was repeated after the death of Emperor Franz Josef in 1916. By marking particular days and not others, theatre owners could also create an alternative public understanding concerning which days had public importance. A coffee shop owner and film exhibitor in the city of Győr, for example, took to celebrating holidays with free film screenings. In 1905 and for the next few years, he celebrated both New Year's Eve and the Jewish holiday Purim by holding free movie marathons that lasted all night.[23] Through their programming, film exhibitors posited theatres as public institutions that took part in the life of the nation.

The political changes after the First World War turned earlier claims about modernity and cinema upside down. By the interwar years, this same argument could only be framed in nationalist terms. In a 1935 debate over government regulation of movie theatre permits in the countryside, István Erdélyi, the manager of a prominent film production company, argued that increasing the number of theatres was crucial, because otherwise 'the isolated countryside' would be cut off from the circulation of national culture and suffer 'cultural stagnation'. Movie theatres provided 'civilization' in remote areas, and could both 'preserve and spread' Hungarian culture.[24]

The meaning of national film: official film policy and the debate on Hungarian cinema

Before the First World War, the government had relatively little regard for the civilizing potential of films and was more concerned with their allegedly detrimental effects on social order and the morality of the lower classes and youth. Authorities were even less enthusiastic about the expansion of the public, which again appeared to pose threats to the social and political order. After the war, the modernizing role of commercial entertainment films gained a different interpretation in conservative and radical nationalist circles seeking to use the persuasive power of cinema to create a new national culture and to form cinema audiences into a national public. If movie theatre owners and liberal intellectuals saw films as bringing civilization and national progress to remote areas, quite the opposite view emerged from conservative critics who believed that commercial films were destroying authentic culture, particularly in the vulnerable countryside.

The Hungarian government had a conflicted approach to commercial films. On the one hand, like most other governments across Europe, it developed a new interest in regulating the film industry. Hungary introduced film censorship for the first time in 1920, and film exhibition was moved from the authority of local governments to that of the Interior Ministry. The state also began to discuss the importance of national film and the idea that entertainment films could be of national significance gained widespread acceptance. Conservative critics viewed cinema audiences as raw material for creating a new, national public and films as a vehicle of national propaganda. Films could represent the nation in a positive light, they could highlight unique national customs and the outstanding cultural achievements of national culture.

In 1925, József Pakots, a conservative member of parliament, argued that 'Hungarian film production is not simply an industrial issue, but an important cultural, social, economic, and ... propagandistic issue as well.'[25] At the opening ceremony of the revamped censorship committee in 1928, the Minister of Interior Affairs, Béla Scitovszky, proclaimed that films exhibited in Hungary should reflect 'national feeling and thought' and serve 'the nation's holiest interest'.[26] Sándor Jeszenszky, an advisor to the Ministry of Culture and Education (and from 1933 one of the senior members of the film censorship committee) argued that the question of Hungarian cinema was one of the most important issues for national culture, because film was 'an excellent tool for bringing into existence a unitary world of feelings and thought, a singular national taste, able to penetrate the entire society'.[27]

From the earliest years of the Horthy regime (1920–44), conservative government officials viewed commercial film and movie exhibition with suspicion, questioning whether commercial and national interests were compatible. In the early 1920s, the

Interior Minister asserted that there was a conflict between commercial and national interest in matters of film exhibition. According to conservatives, those in the film business (movie theatre permit holders and owners) were motivated by profit and not by the best interests of the nation. The entertainment industry and the film industry in particular was understood by the public as being both Jewish and highly profitable, at a time when conservative nationalists blamed Jews for Hungary's predicament; for shirking army duty, war profiteering, 'stabbing Hungary in the back' and for 'leading' the revolutions of 1918 and 1919. In a populist move, the government withdrew cinema permits from previous owners, most of whom were Jewish, claiming that they were 'suspect from a national standpoint' and redistributed them to loyal national subjects reliant on government support, primarily war widows, war veterans and right-wing charitable organizations.

As Justice Minister Vilmos Pál Tomcsányi argued in 1921, movie theatre owners were usually interested only in business profits, and this often conflicted with the national interest:

> I do not want to condemn old movie theatre owners in general. I simply said that they worked for business profits, and the [national] perspectives which I emphasized did not prevail to the proper degree. ... [Since] they were mostly led simply by business profits ... the pieces they exhibited did very little to bring out the desired results in the souls of audiences, the results that this new ordinance wishes to attain.[28]

The new owners were expected to hold screenings of educational, national-themed films once a week and to offer the revenues as a form of tax. As most owners soon found, however, it was more profitable to pay the required sum without screening the poorly attended national films. By the mid-1920s, the new system had collapsed entirely; the new owners, lacking professional experience, were unable to turn a profit and theatres closed; Jewish owners were once again allowed to participate as managers of the theatres. Audiences seemed to prefer the modern commercial films that critics deemed un-Hungarian.

In contrast with liberal commentators who explained these changes as the spreading of civilization to the countryside, nationalists insisted that Budapest was colonizing the rural countryside. Critics easily dismissed Budapest audiences as Jewish or under the influence of Jewish culture. Conservatives were more concerned about rural audiences and were particularly perplexed by the popularity of 'modern' films in the countryside. Although the rural masses were often portrayed as uneducated, with underdeveloped national sensibilities, they were also considered the basis of a new, traditional Christian Hungarian nation. It was, therefore, alarming to see that this authentic national core was becoming part of modern commercial culture as well.

In his 1920 book *Three Generations,* the highly influential conservative historian Gyula Szekfű articulated the common belief that Jews were taking over Hungarian cultural and intellectual life and the media.

The Budapest factory colonized the country. The height of this development was when peasant women wearing headscarves who came to the weekend fairs to sell butter and eggs ... began to enter the bookstore looking for [the popular society magazines featuring stars of the new mass culture] *Szinházi Élet* [*Theatre Life*] and *Érdekes Újság* [*Interesting News*].[29]

For Szekfű, the spread and success of urban mass culture amounted to the destruction of authentic Hungarian culture and national identity. In his view, the rural public seemed to accept these 'Jewish' cultural products as Hungarian and saw no contradiction between modern Budapest and national identity. As a result, it could no longer protect its authentic national culture.

In 1934, the right-wing nationalist film director István György similarly argued that the general success of commercial, urban films in the countryside could be explained not as a new cultural preference, but as an effect of the power of 'capital' on the one hand (distributors looking for profits, and filmmakers trying to make a living), and the failure of audience education on the other. This was a consequence of the 'total absence of a consciousness of Hungarian feeling' in much of the population, together with 'the dictatorial demand of the masses that wish to see these kinds of films'.[30] By claiming that neither the capitalist context of production nor the commercial method of consumption served national culture, György posited commercial practices as being incompatible with national values. Critics suggested that if audiences liked modern commercial films, it was because they had been misled or seduced by Jewish commercial culture.

But when it came to strategies for correcting this, many critics noted that there were simply not enough quality 'national' films available for exhibitors and audiences. Despite frequently repeated demands for 'more Hungarian' films, the government was unprepared to support a new, national film production and relied instead on the market, which in turn failed to deliver the desired films. By the mid-1930s, attacks on Jewish filmmakers became more frequent. Radical nationalists claimed that Jews could not create real Hungarian films because they lacked the authentic spirit required for it. A 1937 propaganda pamphlet by a radical right wing group made the same argument in much cruder terms:

> Hungarian film production is in foreign hands. It serves foreign goals, spreads immorality and filth, and turns Hungarian ideals inside out, disgracing them. ... Székely [one of the most prominent directors of the 1930s] has no connection to the Hungarian race, and therefore is incapable of directing a film about the Hungarian people![31]

Conservative and radical nationalist critics increasingly defined urban, commercial films, and eventually films by Jewish filmmakers, as un-Hungarian. By the end of the 1930s, the government decided that true national cinema would be possible only through the exclusion of Jews from the film industry. The 1938/39 anti-Semitic laws targeted Jewish intellectuals and artists, making it illegal for Jews to play any significant role in the creation of Hungarian films.

But in the minds of radicals, even the removal of Jewish filmmakers failed to solve the problem: the effects of urban commercial culture remained detrimental in rural villages with authentic Hungarian culture. Radical nationalists who considered rural Hungary the last repository of authentic national culture described the modernizing effects of cinema in very particular terms. Conservative politician József Közi Horváth argued in parliament in 1943 that 'Hungarian villages are being corrupted by films'. He demanded that the Minister of Culture 'block the creation of immoral films' by creating a double standard for film distribution with stricter criteria for films played in villages, in order to prevent authentic Hungarian culture from falling victim to the allure of commercial modernity.[32]

Conservative provincial representatives in parliament similarly decried the seemingly irresistible pull that modern films exerted on the rural population. According to conservative critics, films seemed to be transforming traditional culture overnight. Abandoning local tradition, young men and women imitated the dress, speech, behaviour and attitudes of film stars. A 1943 article in the daily paper *Hungarian Sunday* lamented the

> sad changes that have taken place in village life since the introduction of movies. The only things you hear in the streets are film songs ... On the weekends, village girls flock to the cities to get their haircuts and style their hair as certain film actresses do. Before village fairs, it has even happened that young men have gone to the city to get perms as they have seen it on film stars.[33]

Despite government efforts to encourage conservative national pictures, the films that remained popular throughout the period were 'syrupy' stories about modern, urban commercial Hungary and the middle classes. Some of the most successful films of the decade were *Meseautó* (*Dream Car*, Béla Gaál, 1934), a romance featuring a beautiful working girl who falls in love with a car and a bank director who falls in love with her, and *Hyppolit, A Lakáj* (*Hyppolit, The Butler*, István Székely, 1932) about a nouveau riche family and their snobbish butler. Conservatives who saw the film industry as broadcasting an urban, liberal, capitalist perspective through the films were in some respects correct: the films did offer a completely different vision of Hungarian modernity from that favoured by the conservative government elites.

Entertainment films and the popular vision of Hungarian modernity

Although right-wing critics – and some government officials – were interested in creating a national film industry that would produce movies both reflecting and rekindling an authentic national identity, short-lived interwar Hungarian governments lacked the political will and the finances to build one. As a result, the Hungarian film industry remained an almost completely privately funded commercial operation throughout the 1930s. Considering the intense cultural and political interest surrounding Hungarian-made films, it seems paradoxical that until the late 1930s the average annual proportion of Hungarian-made films exhibited in movie theatres was less than 5 per cent. The largest exporters to Hungary were the US and, after 1933, Germany. Nevertheless, Hungarian films still received a great deal of public attention precisely because of the ideological importance given to them as national cultural products, and also because successful Hungarian films (with very few exceptions, Hungarian films were commercially successful) expected an audience of 100,000 viewers, more than any other cultural product.

Before 1938, the film community was a small group of mostly Jewish intellectuals, artists and businessmen and women from the urban middle class. A majority of them had a university education, and a good number of them also travelled and lived abroad in the 1920s and 1930s, learning their trade in Austria, Germany, France, England and the US. A number of film professionals participated in the creation of public culture not only as directors, screenwriters and producers, but also as journalists and writers active in public debates; Lajos Biró, István Székely, and Menyhért Lengyel, for instance, wrote for newspapers as well as films. They belonged to the circle of liberal intellectuals, artists, politicians, writers and entertainers responsible for creating Hungarian popular culture. For them it was very clear that commercial modernity could contribute to national progress, and that Jews were Hungarians. The culture that they created expressed their capitalist, liberal, democratic worldview. Romantic comedy films were the dominant genre of the 1930s. Most of these films had similar casts and were written by a small group of film writers, producing stock stories that imagined urban commercial modernity as the glamorous way of the future. In the films' fantasies, this was the common ideal to which everyone, including traditionally dressed, rural, Christian Hungarians, aspired. This urban commercial modernity appeared not as dangerous, immoral, or deceptive, but rather as the solution to most of the problems of backward, rural society. It was a source of renewal, offering an alternative to the financially, morally, and culturally stagnant countryside.

The view of Hungarian modernization presented in these films was a marginalized alternative that had been banished from political or serious mainstream intellectual debate. But in popular culture, it remained a common way of representing Hungarian modernity until the late 1930s.[34] The conservative nationalist conviction that rural Hungary had a separate, authentic culture untainted by urban modernity was completely alien to these films, in which rural Hungary appeared as already and irreversibly connected to the city and the world beyond by modern channels of communication, culture and transportation. Trains, cars, radio, films, novels, newspapers and magazines, popular music and fashion all featured in the films' representations of the countryside, and contributed to making it part of the modern world. In these films, the Hungarian countryside was represented as a bucolic haven to outsiders who know nothing about it, but as stifling and backward to some of the young people living there.[35] Rather than being a romantic place of pure morals and traditional values, it was part of the urban cultural circuit, somewhat disadvantaged by its distance from the city, and presenting fewer opportunities for fulfilling the dreams and promises of modern culture. It was only through immediate contact with the city, or with a person representing urban culture, that individuals from the countryside could find fulfilment and happiness. The interchange with urban culture was necessary for initiating change.

The romantic comedies affirmed that people from the countryside were not essentially different and could quickly learn to lead modern lives. They also insisted that modern attitudes and values promised individual freedom, self-fulfilment and happiness. Young women as well as men could transform their own lives by imitating their role models on film. In *Köszönöm Hogy Elgázolt* (*Thanks For Running Me Over*, Emil Martonffi, 1935), for instance, a young girl living in a village is constantly complaining about her boring life. Longing to live the life of independent modern girls, she reads romance novels, orders the latest jazz music from the city and plays it on her piano. When she gets her first chance to meet a modern young man, she chases after him, boldly imitating the modern independent women she has read about. Ultimately, she finds happiness and a husband and moves to Budapest. Modern identities were presented as immediately appealing to everyone, and often as offering a way out of the constraints of outdated traditional morality and customs.

Conservative calls to renew the national community by returning to the authentic roots of national culture found a somewhat bizarre response in the films. A prime example of this intriguing fantasy is *Iza Néni* (*Miss Iza*, István Székely, 1933) based on an idea by the legendary prima donna Sári Fedák.[36] *Iza Néni* is the story of a female provincial language teacher, passionate about Hungarian literature and culture, who is forced to leave her position after the First World War when her village becomes part of Czechoslovakia. Miss Iza moves to Budapest and begins to work in a nightclub because she is hard-pressed for money. Later, she moves to Paris, becoming an extremely successful bar singer at the *Étoile Noire* nightclub. Yet success changes neither her interest in national progress nor her dedication to her country: she sends all her savings back home and continues to help those left behind in a variety of other ways, including helping a former student to find a job and a wife.[37]

Instead of presenting urban commercial culture and nightlife as destructive of traditional national culture and the countryside, *Iza Néni* represents a bar singer as the upholder of classic Hungarian culture and tradition. The film insists that successful Parisian nightclub stars can also be frugal middle-class employees who save and invest their earnings, providing huge amounts for charitable purposes. A nightclub singer becomes a nationalist fairy godmother, saving a rural village left behind by redefined national borders.

In these films, the rural countryside stood for traditional Hungary waiting to join capitalist middle-class culture. The rural was used as a setting for re-creating the nation in a way very different from that imagined by conservative and radical nationalists. It was not exclusive (even Jewish bourgeois capitalist characters can find their home in the countryside and help create its new culture), and it contained no essential authentic truth. Commercial culture also gained a completely new significance, as a liberator from tradition, a source of independence and an opportunity for self-determination for individuals, and as a creative way of saving Hungarian tradition. Filmmakers imagined a public that could appreciate stories about the success of independent, enterprising individuals from the middle and lower classes who would remake the world according to democratic individualistic values.

Hungarian film narratives and themes often mirrored the fears of conservative nationalists about the effects of commercial modernity on the traditional countryside. Both national traditional culture and commercial modernity gained a more positive interpretation in these popular fables. From the 1920s, conventional political discourse was defined by the conservative, nationalist vision of Hungary as a traditional, rural and Christian nation, but debates over the desirable direction of national modernity also took place in and through cinema. In the arts and commercial popular culture, a very different idea of Hungarian modernity continued to exist throughout the 1930s, and film was perhaps the most influential medium in which a liberal vision of Hungarian modernity prevailed.

Although the vision of modernity that we see in films had no political expression, it was far from marginal in the cultural sphere: it was widely celebrated in popular entertainment. This seeming contradiction was not unique to Hungary; as studies of Nazi cinema in Third Reich Germany have shown, there was a complicated relationship between consumer culture and Nazi ideology, and the desires of modern consumers clashed with Nazi attempts to use entertainment films for propaganda purposes, often producing ambiguous results.[38] The Hungarian situation was exceptional in that until the late 1930s it was largely Jewish filmmakers (writers, directors and actors) who produced these fantasies. Without viable political representation, however, the liberal commercial worldview of the films did not have a significant influence on public debates. While debates over films and film audiences in Hungary were shaped by the political understanding of what modernity meant, filmmakers, exhibitors and audiences were not merely passive subjects of these debates, but also their active agents. Their behaviour and representations suggest a diversity of opinions that challenged the dominant conservative national identity that has come to define our understanding of interwar Hungary.

Notes

1 M. Hansen, 'Fallen Women, Rising Stars, New Horizons', *Film Quarterly* 54, 2000, 10–22. M. Hansen, *Babel and Babylon*, Cambridge, MA: Harvard University Press, 1991. L. Charney and V. Schwartz (eds) *Cinema and the Invention of Modern Life,* Berkeley: University of California Press, 1996. H. Jubin, *Projecting a Nation: Chinese National Cinema Before 1949*, Hong Kong, London: University of Hong Kong, 2003. M. Pomerance (ed.) *Cinema and Modernity*, New Brunswick, NJ: Rutgers University Press, 2006. G. Branston, *Cinema and Cultural Modernity*, Buckingham: Open University Press, 2000. G. Branston (ed.) *Reinventing Film Studies*, London: Arnold, 2000. A. Ligensa and K. Kreimeier (eds) *Film 1900: Technology, perception, culture*, Bloomington, IN: John Libbey Publishing, 2009. M. Jancovich, L. Faire and S. Stubbings (eds) *The Place of the Audience: Cultural Geographies of Film Consumption*, London: British Film Institute, 2003. B. Singer, *Melodrama and Modernity: Early Sensational Cinema and its Contexts*, New York: Columbia University Press, 2001.

2 See for example, the essays in K.H. Fuller-Seeley (ed.) *Hollywood in the Neighborhood: Historical Case Studies in Local Moviegoing,* Berkeley: University of California Press, 2008.

3 S. Szerdahelyi, *Budapesti ujságirók almanachja*, Budapest, 1908, p. 505. Quoted in B. Fabó, 'A Moziépitészet És a Város' [Cinema Architecture and the City], *Budapesti Negyed* 5, 1997, 196.

4 Quoted in B. Fabó, 'A Moziépitészet és a Város', 196.
5 According to government statistics seating capacity increased to 39,167 in 1928 (87 theatres). The number of movie theatres in Budapest actually remained almost constant (86 in 1938, 89 in 1939) partly because the Hungarian government had a strict quota on movie theatre licences in the interwar decades. 'Lajos Illyefalvi, Tiz Ev Budapest Életéből a Világháború Után', *Magyar Statisztikai Szemle* 8, 1929, 217.
6 Henrik Castiglione, *Filmkultura III* 12, 3–5.
7 According to the film historian and theorist Béla Balázs, a lot of these travelling cinemas started out in Calais or Hamburg, ending up in Jassi or Chernovitz, stopping at the market squares of mid-size and larger towns, and visiting small towns and villages only in the rarest of cases. B. Balázs, 'Tíz Évvel Lumiere Után', *Filmspiral* 5, 1999, 45.
8 G. Darvas, 'Magyarország Mozgóképüzemei 1928-Ban', *Hungarian Statistical Review* 8, 1929. Reprinted in Z. Kōháti, *A Magyar Film Olvasókönyve*, Budapest: Magyar Nemzeti Filmarchivum, 2001, pp. 510–18.
9 Ibid., p. 195. According to Henrik Castiglione, a producer and expert on film statistics, in 1935 there were 440 movie theatres in Hungary, 75 in Budapest, 24 around Budapest, 21 more in nine major cities, with the remaining 310 in smaller towns. By 1935 75 per cent of theatres were in the countryside, but only 60 per cent (310 movie theatres) were in small towns.
10 Interwar conservative governments were fearful of political and social unrest from the rural poor; according to the voting system set up in 1919, rural constituencies voted by open ballots. The banning of movie theatre licences had to do both with the detrimental effects films were considered to have on individual morality and with the ban on public assembly.
11 In 1921, 91 of Hungary's 367 movie theatres were in Budapest, 67 in the area around the capital, and 209 in the countryside. These numbers are based on police records of the number of permits issued each year. In reality a slightly (and sometimes considerably) lower number of theatres were in operation: getting a permit was difficult so those who had managed to obtain one often chose to renew it even if they were temporarily unable to operate their business rather than risk lose the permits permanently.
12 L. Illyefalvi, *10 Év Budrapest Életéből a Világháború Után*, Budapest: Budapesti Székesfővárosi Statisztikai Hivatal, 1930, p. 195. According to Henrik Castiglione, in 1935 there were 440 movie theatres in Hungary, 75 in Budapest, 24 around Budapest, 21 more in 9 major cities, with the remaining 310 in smaller towns. Castiglione, *Filmkultura III* 195.
13 I. Romsics, *Magyarország Története a XX. Században*, Budapest: Osiris, 2004, p. 219.
14 See M. Sebōk, *Sokszinū Kapitalizmus*, Budapest: HVG, 2004, and the recent work of Károly Halmos, Gábor Gyáni, György Kövér and András Gerō. For an overview, see *Social History of Hungary from the Reform Era to the End of the 20th Century*, Boulder, CO: Social Science Monographs, 2004.
15 Contemporaries already recognized this as a problem, and thinkers as divergent as the Marxist Erik Molnár, Ferenc Erdei, Oszkár Jászi, and the Jewish liberal intellectual Pál Ignotus agreed. As the latter argued in 1917, conservative critics from the turn of the century generally discussed the problem of modernization as the 'Jewish question' and attributed everything seemingly threatening or problematic about modernization to 'foreign' or 'Jewish' influence. See for example, P. Ignotus, 'A magyar kultúra és a nemzetiségek', *Nyugat* 1, 1908. Also F. Erdei, 'A Magyar Társadalom a Két Háború Között', *Valóság* 5, 1976, 40. (This was written in the interwar years but published only later in this form.) According to Erdei, the reason that Jews could succeed in capitalism was their marginal position as outsiders in a feudal order: 'Capitalism created its own bourgeois society in Hungary, made partly of foreigners, partly out of the less closely tied elements of Hungarian society, Jews and Germans.' For a contemporary analysis, see P. Hanák (ed.) *Zsidókérdés Asszimiláció Antiszemitizmus Tanulmányok a Zsidókérdésről a Huszadik Századi Magyarországon* [*The Jewish Question, Assimilation and Antisemitism: Studies on the Jewish question in 20th century Hungary*], Budapest: 1984, pp. 76–77. A. Gero, *Jewish Criterion in Hungary*, Boulder, CO: Social Science Monographs, 2007.

16 Ignotus in *Nyugat* (1908). Quoted in J. Frigyesi, *Béla Bartók and Turn of the Century Budapest*, Berkeley: University of California Press, 1998, p. 83.

17 The most influential example of this type of argument was the extremely popular tome of conservative historian Gyula Szekfű in *Három Nemzedék és Ami Utána Következik [Three Generations and what came after]*, Budapest: 1920.

18 N.A., 'Villamos Színház' [Electrical Theatre], *A Város* (2 February 1907).

19 J. Hárs, *A Filmvetítés Története Győrött 1896–1912 [The History of Film Screening in Győr, 1896–1912]*, Győr: Győr Megyei Város Tanácsa,1989, p. 110.

20 Ibid.

21 Ibid., p. 109.

22 C. Nagy, *Kanizsai Mozitörténet [Cinema History of Nagykanizsa]*, Budapest: Magyar Filmintézet, 1996, p. 3.

23 J. Hárs, *A Filmvetítés Története Győrött 1896–1912, p. 109.*

24 I. Erdélyi, 'A Szünetelő Vidéki Mozgószínházak Megnyitásának Kérdése' [On Reopening Temporarily Closed Rural Theatres], *Filmkultúra* 7, 1935, 5.

25 J. Pakots, 'Interjection to the Hungarian Royal Government on the delay in implementing government decree number 4503/1925 M.E. on the development of film production', 21 October 1925. 451th session, *Nemzetgyűlési Napló* 35, 1922–26, pp. 103–5.

26 According to Scitovszky, 'The guiding principles defined in the law provide merely the frame, to be filled with meaning and life by the demands of national feeling and thought, demands of the ethical commands of the nation's holiest interests, demands of social morality and of good taste. ... We all know that the national interest and ethics, the national and social moral order demands the enforcement of considerations so wide-ranging and far-reaching that cannot possibly be enlisted in austere paragraphs. But I am convinced that the members of the committee, by their moral and intellectual qualities, will enforce in their work with the right subjective considerations even principles that cannot be contained in written laws.' 'Nation-building principles in film censorship', *Rendőr* II, 3 (21 January 1928) 5.

27 S. Jeszenszky, 'A Magyar Film-a Magyar Kultúra Egyik Létkérdése' [Hungarian Film Is One of the Most Important Questions of Hungarian Culture], *Filmkultúra* 7, 1934, 1.

28 Nemzetgyűlés 170. ülése. *Nemzetgyűlési Napló* 9, 1920–22, 121–42.

29 Gy. Szekfű, *Három Nemzedék És Ami Utána Következik*, Budapest: ÁKV-Maecenas, 1989, p. 37. The film has been lost, the summary is provided by Gyöngyi Balogh in B. Varga (ed.) *Játékfilmek Hungarian Feature Films 1931–1998*, Budapest: Magyar Filmintézet, 1999, p. 25.

30 I. György, 'Magyar Film', *Nyugat* 27, 1934.

31 Quoted in T. Sándor, *Örségváltás Után*, Budapest: Magyar Filmintézet, 1997, p. 93.

32 J. Közi Horváth, 'Hungarian Film before the Law', *Magyar Film [Hungarian Film]* 1943, Dec. 22, 2.

33 Quoted in Sándor, *Örségváltás Után*, p. 202.

34 That this perspective has been largely forgotten has much to do with the fact that after the Second World War and in socialist Hungary, the liberal-Jewish worldview represented in these entertainment films was rejected as 'bourgeois' and 'capitalist' and the films were banned. (The official explanation was that many of the actors and actresses had emigrated illegally to the West. Many left earlier fleeing fascist persecution.)

35 *Az Új Rokon (The New Relative*, Béla Gaál, Harmónia Film, 1934) is about a rural family whose sour mood and traditional conservatism is lifted by a newly discovered modern young American relative. *Köszönöm Hogy Elgázolt (Thank You For Running Me Over*, Emil Martonffi, Hermes Film, 1935), in which a young girl longs to escape village life and become like the heroines of her popular novels. *Kölcsönkért Kastély (The Borrowed Castle*, Ladislao Vajda, Harmónia Film; 1937), in which a young girl living on a country estate with her conservative mother goes to the city and is able to realize her dreams and marries a man without status.

36 In his memoirs, Székely claimed that he tried to neutralize the film's political message of irredentism, an obsession of the actress and producer Sári Fedák, as much as possible. I. Székely, *A Hyppolittól a Lila Akácig*, Budapest: Gondolat, 1978, pp. 104–9.
37 The film has been lost, the summary is provided by Gyöngyi Balogh in G. Balogh *et al.* (eds) *Játékfilmek Hungarian Feature Films 1931–1998*, Budapest: Magyar Filmintézet, 1999, p. 25.
38 J. Bruns, *Nazi Cinema's New Women*, Cambridge, New York: Cambridge University Press, 2009. A. Ascheid, *Hitler's Heroines*, Philadelphia, PA: Temple University Press, 2003.

6

THE CINEMATIC SHAPES OF THE SOCIALIST MODERNITY PROGRAMME

Ideological and economic parameters of cinema distribution in the Czech Lands, 1948–70[1]

Pavel Skopal

> We have nationalized not only film production … [but] also … film renting, which means that it is impossible for foreign film agencies to nose around here freely and sell all kinds of rubbish as they did in the past. Film distribution is fully in the hands of the state, which has to ensure that even the most expensive films are shown everywhere and not only in those places where the rental fee has been paid.[2]

This statement, made by the Minister of Information Václav Kopecký at the Eighth Congress of the Czechoslovakian Communist Party (KSČ) in March 1946 (two years before the communist putsch in February 1948), underlines how the party perceived cinema.[3] For the KSČ film distribution was seen as a means of creating a new national film culture and a new socialist man.[4] The process of cleaning out the 'rubbish' was not supposed to be limited by commercial considerations. On one hand, the leftist vision of a new culture built up after the Second World War was widespread in Czechoslovak cultural circles and reached far beyond the limits of Communist Party members; on the other hand, the Party promoted the most harsh attacks against popular generic film production labelled as 'kitsch' or 'trash'.[5] The immediate subordination of film distribution practices to the goals of the KSČ's 'cultural policy' was not possible, however, because of two interconnected constraints: the financial demands of operating the nationalized domestic film industry combined with the inertia of existing distribution mechanisms and contractual obligations with foreign distributors to delay the implementation of the 'cleaning' until almost two years after the communist putsch. The subordination of distribution practices to ideological goals was conditioned by financial criteria connecting the economic outcomes of distribution with the requirements of the state budget. The definition of a suitable balance between the 'cultural-political and economic views' of cinema became the central, explicitly stated criterion for the management of distribution practices during the whole period of 1948–70.

The Czechoslovak film industry was nationalized in August 1945.[6] The KSČ won the largest share of the votes (31 per cent) in the post-war parliamentary election, but it was the government crisis in February 1948 that allowed the communists to take power and form a new government. The purges that followed the putsch affected the film industry, even though it was already under the KSČ's influence: over 200 employees were dismissed from their posts. In the autumn of 1948, the Party announced that it was taking a sharp political line against the 'reactionaries', and this position determined film policy for the next five years. A partial freeing of the cultural sphere in the years of the 'New Course' from 1953 to 1956 was followed by a conservative reaction in the two years preceding the announcement of the achievement of socialism in 1960.[7] The ensuing years witnessed a period of gradual liberalization (the conservatives' position was weakened by the 22nd Congress of the Soviet Communist Party), culminating in the 1968 Prague Spring but terminated by the Warsaw Pact invasion of Czechoslovakia in August 1968. The process of 'normalization' that followed was a counter-reaction by the conservative wing within the Communist Party. In this chapter I shall discuss a series of questions about the period from 1948 to 1970, such as: what was the organization and function of cinema distribution? How was distribution moulded by the ideological task imposed on it by the power centre? In considering these issues, it is important to keep in mind that cinema distribution served not only as a means of disseminating propaganda and ideologically correct entertainment. The distribution system had its own propagandistic role, through the alleged ability of the regime to serve the people by presenting valuable works of art, in both the cities and the countryside.

Socialist modernity and the programme of the cultural revolution

As Katherine Pence and Paul Betts argue, post-colonial studies have given impetus to discussions of 'multiple' or 'alternative' modernities, liberating the concept from its exclusive identification with the West.[8] The debate over what kind of modernity or process of modernization could be ascribed to the Soviet Union of the Stalinist era goes back to the 1950s, however.[9] Some recent contributions argue that although discussions of totalitarian regimes make reference to their anti-modern elements, it is more productive to understand them as specific versions of modernity. Shmuel N. Eisenstadt identified the modernities of the Soviet and fascist types as the first distinct 'alternative' versions of modernity, set within the framework of the Enlightenment and major historical revolutions. The strongly anti-Western rhetoric of communist regimes does not imply that the socialist model was anti-modern.[10] As David L. Hoffmann has argued, socialist regimes surely drew on distinctively anti-modern topics, but they used them for modern mobilizing purposes.[11] As well as displaying many of the destructive elements of modernization, the stress that socialist societies placed on industrialization, secularization and universal education represented an alternative version of modernization rather than a revolt against modernity.[12] Besides, the contemporary discussion points out that the concept of totalitarianism does not fit the socialist countries in the post-Stalinist era and tries to find an apt concept representing the tensions and contradictions in the character of these regimes.[13]

An historical analysis of the role played by cinema in the socialist version of modernization would be a useful contribution to the discussion. The emblematic case might be the accent put on the process of 'cinefication': the provision of film screenings throughout the whole state, primarily for educational and propagandist purposes. The term originates in the USSR in the 1920s when, along with 'electrification' and 'radiofication', it established a vocabulary intended to indicate the promises of revolutionary change through modern technologies.[14] It was a term widely adopted and used by the film industry and film journals in Czechoslovakia after the Second World War. Despite the Czech lands having had a dense network of cinemas since the late 1920s, the 'cinefication' process was a proposed goal for the nationalized industry. After the communist putsch its application became more widespread: from the cinefication of the countryside by permanent venues and travelling cinemas in small villages, through 'enterprise cinemas' in factories, to 'cinemas for enlightenment' run by educational organizations.[15] The regime regarded the cinema network as an important asset in propagating the politics of enlightenment (for example, in the annually repeated campaign 'Spring in the countryside' launched in 1951), and strengthening the distribution network became an important goal of the planned post-war development. The first five-year plan assigned the Czechoslovakian State Film the task of increasing the number of theatres, spreading them equally across the whole country and reaching the level of one seat for every ten citizens.[16]

Industrial, administrative and political agents intended to use distribution mechanisms for various goals, and the main contradiction in their use was explicitly reflected in the frequently articulated opposition of 'ideology versus economy'. For a deeper understanding of the discursive construction around this conflict, it is helpful to reframe the topic through the modernization thesis. The concept of 'the cultural revolution' served as a programmatic articulation of socialist modernity.[17] One of the main goals of the KSČ cultural revolution highlighted at KSČ party congresses in 1949, 1958 and 1959 was 'to instill the approach to the world, the new worldview, the new ideology, the new thinking, the new culture, the new morality into all the citizens of Czechoslovakia'.[18] The creation of the 'new socialist person' was part of a programme that was distinctively 'modern', if we understand modernity 'in terms of two features common to all modern political systems: social interventionism and mass politics'.[19] After the 'achievement of socialism' in 1960, an official periodization of the cultural revolution was established: a preparatory period (until 1948), the first period (from 1948 to 1958) and the second period of the real revolution, after the 11th Party Congress in 1958. The concept of the cultural revolution was permeated by claims of modernization: the cultural revolution was 'mingled with the technical revolution' and the typical feature of its second stage was 'a fight for the use of science in the production as well as in the leadership of society, a fight for a new technique'.[20]

In his speech at the conference on film distribution in 1958, the Minister of Education and Culture, František Kahuda, summarized the role of cinema in the first period of the cultural revolution: film had a leading role because of its mass character, popularity and comprehensibility for all sections of the population.[21] The years 1958–60 were marked by an anti-liberal turn in cultural policy, followed by an announcement of the completion of socialism in 1960. In the reaction to the political events of the second half of the 1950s (the uprising in Hungary, disorder in Poland, the Berlin crisis, a new rift with Yugoslavia), the conservative wing in the KSČ attacked alleged 'revisionists', and this reaction badly affected the cultural sphere, including cinema production and distribution. After 1960, however, a slow process of liberalization in the cultural sphere begun. This is illustrated by the increasing number of Western movies screened, by comparison to a declining number of Soviet ones (see Figures 6.1 and 6.4 at the end of this chapter). In film distribution, the declaration of the new stage of socialist modernization paradoxically created more space for images from the West, mainly in the form of American and West European movies. The distribution of Delbert Mann's movie *Middle of the Night* (USA, 1959) provides an illuminating example of how the values of socialist modernity intermingled with images of Western modernity and consumerism, which were perceived as a threat even in the critical version offered in the Mann movie. In 1959, the board of censors objected that the film was damaging in its topic, because 'it indirectly emphasizes a high standard of housing and domestic life of an ordinary American family and the generosity of a wealthy businessman who married a poor girl'.[22] Five years after being banned, however, it was released and attended by half a million people.

Prelude: economic arguments in the service of political objectives (1945–48)

Besides the technical condition of the cinemas, the main problem during the postwar period was the lack of suitable films. Initially, only unacceptable German pictures and Czech films with German subtitles were available.[23] The situation was solved by import contracts – mainly with Hollywood studios and Sojuzintorgkino.[24] The contract with the Soviet film industry signed in July 1945 favoured Soviet productions and granted Sojuzintorgkino 60 per cent of the screenings in cinemas and the purchase of at least a hundred Soviet films in the first year of the contract, with that number increasing by 5 per cent a year. The Soviet film industry was, however, unable to meet such volumes of production, and the shortage of movies for distribution was eventually solved by an agreement with the Motion Picture Export Association (MPEA), which ensured that 34 American films were accepted for distribution in 1946, and 82 in 1947.[25] There were almost as many screenings of movies from the USA and Western Europe in 1946 as there were screenings of Soviet productions; in 1947 Western movies dominated the screens with an almost 50 per cent share of all screenings, and 55.6 per cent of all attendance.

Distribution practices favoured Soviet movies, however. In his speech at the meeting of regional cinema directors on 12 February 1947, the General Director of the Czechoslovakian Film Association, Lubomír Linhart, explained that these practices resulted from the economically advantageous contract with Sojuzintorgkino, which divided rental income equally between the Association and Sojuzintorgkino, whereas in the MPEA contract the Czech distributor only retained 35 per cent of the rental.[26] According to Linhart cinemas represented the most important profit segment of the nationalized film industry, and distribution had to abide by criteria of 'profitability'. Czechoslovakian films, for which no profit shares were returned abroad, should therefore receive the most preferential treatment, followed by Soviet and Swiss films provided under advantageous contracts. Although it was preferable to schedule screenings of these films for the peak viewing times between Fridays and Mondays, Linhart complained that it was American films that were most often scheduled for weekends.

Linhart's assertion intentionally discounted the issue of popularity and the higher attendance achieved by Western films. He presented viewing behaviour as involving a 'visit to the cinema', not to a particular film; since this was something that mainly occurred at weekends, it was possible to increase profitability by means of mechanical changes in scheduling. The possibility of educating viewers in the right cultural and political values by sanitising the films on offer was already explicit: 'Mercilessly eliminate, within the shortest time possible, the loss-making foreign films: not only all the loss-making films from the economic point of view, but also those films that are loss-making from the cultural viewpoint of the development of the cultural education of Czech viewers.'[27] Linhart's argument was repeated in the Party's correspondence, and pressure was put on the distribution system to implement the objectives of the proposed measures.[28] The transfer of Soviet films to the more advantageous weekend dates was justified as a way of achieving higher profits.

The Communist Party presented itself explicitly as the creators of the ideology of the new Czechoslovakia. As Minister Kopecký's speech implied, cinemas had an important place in communist rhetoric: its nationalization was a symbol of the victory of the 'national and democratic revolution' over the old order; and the new distribution system was portrayed as a matter of public interest and as a democratic and educational tool not restricted by profitability. Both the sanitised distribution programme and the egalitarian distribution system itself implied a distinctively 'modern' post-war order.

Unsuccessful attempts at creating a new type of national film culture (1948–50)

After the February 1948 putsch, the change of regime was not immediately apparent in cinemas or in the distribution system. In the autumn of 1948, however, Distribution Director Jaroslav Málek defined the task of distribution as the 'systematic spreading of film work among the masses'. In the fulfilment of this task, regional directors and cinema managers had to 'help those films that are valuable from cultural and political points of view' by promoting and organizing mass attendance. On the other hand, 'escapist' films were labelled as harmful and attendance should not be encouraged for 'those films that are the inescapable evil that we ... still need to deal with'.[29] A year and a half later this task was confirmed by the Party's Central Committee (ÚV KSČ), which presented a binding guideline for Czechoslovakian State Film (CSF). It required distribution practices that 'educate viewers especially by the means of our new films as well as by the Soviet films and the films from people's democracies'. Mass organizations, public education and school administrations were to provide mass attendance.[30] In order to meet the objectives of educating viewers by directing them towards ideologically suitable works and purging the remaining relics of the pre-putsch cultural policy from distribution, two crucial measures were taken. At the beginning of 1949 so called 'circular distribution' began, ensuring the fast circulation of 'films that are most significant from the distributional point of view' in the 'circuit cinemas'. These films were put into 13-week programme sequences. Each sequence had to include six Czechoslovakian and four Soviet films, leaving three places for films chosen by the Distribution Centre. This model, which only worked until the second half of 1953, represented the highest degree of centralization and control.[31]

The number of new films screened had fallen sharply from 199 in 1947 to 92 in 1949 and 74 in 1950, and attendance had dropped from 122.2 million in 1949 to 98.8 million in 1950. The distribution system was given the task of significantly changing viewers' motivation by educating them in a new 'attendance tradition' and teaching them, in the words of Jaroslav Málek,

> not only to attend the new films but to learn from progressive films again and again, and to change the current state in which a viewer always sees a film only once […] the film will really become a possession of the working class, who will protect it against the reactionary danger of embarrassingly low attendance by enjoying these films and learning from them in the cinema on a large scale.[32]

The model of a new viewer and a new film culture was to be based on the activity of visiting cinemas as a political act of self-education. Seeing Soviet films was to be a 'manifestation of inseparable friendship with the Soviet Union', so it was not important if the viewer had already seen the film.[33] Avoiding Soviet films was 'a matter of dubious personal prestige of enemies of the time'.[34]

To motivate cinema employees to ensure attendance at 'valuable films', a system of financial bonuses was created for cinema managers. Attendance competitions were announced, and when 'important' Soviet films, such as *The Fall of Berlin* (*Padenije Berlina*, Michail Čiaureli, SU, 1949) or *Secret Mission* (*Sekretnaja Missija*, Michail Romm, SU, 1950), were being shown, managers were ordered to programme only films that would not attract viewers' attention in competition with them.[35] It was important that such administrative and institutional mechanisms were created in order to guarantee sufficient attendance. The task of a regional promotion official was to 'educate viewers and increase attendance'; 82 per cent of the promotion costs in 1949 were spent on 'progressive films', i.e. films from Czechoslovakia, the Soviet Union and from other 'people's democracies'.[36] These mechanisms represented a new level of insecurity, and some managers complained that even Communist Party officials did not go to see the progressive films.[37]

The significant decrease in attendance showed that efforts to reach the objective had not been successful – despite organized attendance, the average number of viewers per screening at a Soviet film was 148, while for 'other' non-socialist productions it was 228 (see Figure 6.3 at the end of this chapter). Attendance decreased despite the fact that mass visits to theatres were organized and cinema tickets were sold to schools and factories, and ticket sales were included in the viewing statistics despite the fact that large numbers of the tickets were not used. An economic expression of the failure to create a new, 'ideal' film culture was a decrease in revenues from CZK 389 million in 1948 to 245 million in 1950. In a reaction to the situation, Distribution Director Málek promised a greater variety of topics, hoping to bring 'escapist' film viewers back to the cinema and persuade them of the importance of progressive films. In the years of the toughest Stalinism, economic demands and the failure to change viewers' tastes and values created a slight opening in the distribution system for films that were not compatible with the state's cultural programme.

Conflicts between the Czechoslovak State Film and state authorities (1951–56)

In 1951 CSF took various measures to increase revenues and differentiate the range of films in distribution. 'Colour epic films', particularly the Austrian ice-revue movie *Spring on the Ice* (*Frühling auf dem Eis*, Georg Jacoby, Austria, 1951) and the Czech hit with the immensely popular actor Jan Werich in a double-role *Emperor's Baker – Baker's Emperor* (*Císařův pekař – Pekařův císař*, Martin Frič, Czechoslovakia, 1951), were exhibited for an 'increased flat fee', with all tickets sold at the highest price set for the given cinema category. In the next two years, other foreign films were shown in this way: *Child of the Danube* (*Kind der Donau*, Georg Jacoby, Austria, 1950), *The Tiger's Claw* (*Der Tiger Akbar,* Harry Piel, West Germany, 1951), *One Summer of Happiness* (*Hon dansade en sommar,* Arne Mattsson, Sweden, 1951) and *To the Eyes of Memory* (*Aux yeux du souvenir*, Jean Delannoy, France, 1948). In 1953 the practice of showing films at an increased flat fee came to an end and a larger number of films were shown within an 'extended programme' for so-called 'normal fee increased by 1 CZK'.[38] Sixteen films were shown in this kind of extended programme in 1953, including three ideologically burdened but very popular Czech films: *Africa I* (*Afrika I – Z Maroka na Kilimandžáro*, Miroslav Zikmund – Jiří Hanzelka, 1952), *Angel in The Mountains* (*Anděl na Horách*, Bořivoj Zeman, 1955) and *Rafter's Card Game* (*Plavecký mariáš*, Václav Wasserman, 1952). The other movies came from Western countries, among them *Fanfan the Tulip* (*Fan-Fan la Tulipe,* Christian-Jaque, Italy/France, 1952), *Where No Vultures Fly* (Harry Watt, GB, 1951) and *The Beauty of the Devil* (*La Beauté du Diable*, René Clair, France/Italy, 1950). CSF was, however, heavily criticized by the Ministry of Culture, the State Planning Office and the Central Government Presidium for violating state discipline by showing these films at a surcharge. The extended programmes were cancelled and 'thoroughly investigated'. The State Film defended itself by claiming that it had 'tried to add culturally and politically valuable film supplements (often Soviet documentaries that would normally not be seen by most viewers) to the ideologically weaker films'.[39]

The practice of extended programmes was an obvious attempt to solve the incompatibility of ideological and economic demands through distribution rather than production. The centrally directed and ideologically controlled programme of production had almost collapsed: while the five-year plan counted on producing 52 movies a year, only seven features were completed in 1951. Although it did not succeed in changing viewers' preferences, the cultural programme of cleaning the cultural sphere was still implemented. The extended programmes 'reconciled' the contradictions with a hybrid of Western-produced, entertaining pieces like *Fanfan the Tulip* and propagandistic Soviet documentaries presenting the alleged results of the successful socialist modernization: according to a contemporary article, the documentaries showed Armenian children, whose fathers had had no opportunity to learn reading and writing, enjoying modern freeways, energy-producing dams, new species of cattle bred by the Soviet scientists, and a children's railway.[40] The entertainment values of the few popular pictures, both indigenous and imported, were framed by the emblematic images of modernized society: industrialization, science, education, transport infrastructure and leisure time.[41]

Decentralized distribution and restored ideological control (1957–60)

The number of new feature films in the category of non-socialist countries doubled from 23 in 1955 to 52 in 1956. Attendance rose from 129 to 146 million (see Figures 6.4 and 6.2 at the end of this chapter). The demands of the state financial plan put pressure on the system and motivated more frequent showings of attractive, but ideologically problematic Western produced films. The increasing incongruity between the priorities of the centrally planned enterprises and Communist Party authorities headed towards conflict.

In 1957, the distribution system was extensively decentralized, giving two new players in the system economic interests in distribution.[42] National committees became cinema owners, with cinema revenues representing an important part of their budgets. At the same time, 'cost-accounting' (*khozraschet*) Regional Film Enterprises became the central link between the National Committees and CSF. As a result, programming started to take place on multiple levels; it became subject to stricter supervision and criticism by CSF and especially by the government and Communist Party authorities.

As part of the decentralization process, the Central Film Rental Office (CFRO) was established on 1 January 1957. Three months later cinemas were transferred to municipal and local national committees.[43] The new system soon required a correction of distribution practices in line with the criteria set by the Communist Party. In September 1957 the Czechoslovakian Communist Party Central Committee (ÚV KSČ) presented a report on the 'current cultural and political problems of film distribution'. Blaming the rapid increase in attendance at Western films and the declining popularity of Czech and Soviet films on weakened supervision by regional committees and Party structures, the report commented that 'Regional film corporations are trying to meet their difficult financial plans by increasing the number of screenings of supplementary Western films.' Nevertheless, the financial plans were not met. Based on the report, a resolution of the ÚV KSČ Secretariat instructed the

Minister of Education and Culture, František Kahuda, to enforce the new rule of not presenting more than 35 per cent of Western movies.[44]

Decentralization sharpened the conflict of interests in the distribution system between the ideological requirements of the Party apparatus and the efforts of the national committees and the regional film corporations to raise additional revenue and meet their financial targets. The 35 per cent limit became the basic regulatory instrument and the central criterion of distribution practice for the following years; it not only limited the number of films, but until 1964 it was in force for the number of screenings and of viewers as well.[45]

The brief period of weaker control over distribution in 1957 coincided with the highest attendance rates in Czech cinemas. In the following three years, the share of capitalist productions in the total screenings decreased, while the share of Soviet films grew rapidly (see Figure 6.1) – in effect, the overall attendance decreased (Figure 6.2). The effort to stop the short-term liberalization and diversification of films in distribution was reinforced by Minister Kahuda in a speech at the National Conference on Film Distribution in October 1958, in which he declared that the task was to 'cleanse cinema programming of rubbish' and 'increase the ideological effectiveness and educational effect of the film'.[46] Although extended programmes were supposed to be abandoned, in the end they survived, partly because a similar form was planned in the Soviet Union (called the Grand Film Programme in the Czech press). Another explicit measure was a reduction in the number of screenings, justified by the argument that the number of unprofitable screenings had to be reduced because they raised wage costs.[47] In reality, the reduction only affected high-attendance Western productions, while the number of screenings of Soviet films kept growing. Governed by injunction of the 11th KSČ Congress to 'complete the cultural revolution', the increased ideological restrictions imposed on distribution in the three years after 1957 should be seen as part of the preparations for declaring a successful transition to socialism, and, above all, as part of the campaign against 'revisionism'.[48] As the regime replaced the ritualized claim of attained socialism with a tendency to pay more attention to consumerist values, the 1960s saw a gradual relaxing of these restrictions, evidenced by the increasing number of Western films in circulation (see Figures 6.1 and 6.4).

From the 'completion of socialism' to the economic reforms – and back again to the communist 'normality' (1961–70)

Even if no fundamental structural changes occurred in the 1960s, the decade was characterized by a gradual and limited liberalization of film distribution, accompanied by growing receipts. This trend culminated in 1969, when receipts exceeded those of 1957. Although overall attendance declined throughout the period (with the exception of 1969), the growth in cinema receipts resulted from increasing admission prices for extended programmes, double screenings, wide-screen films, and open-air 'summer' screenings.

The statistics for the Czech part of the Czechoslovak state suggest a quite similar linear trend. The number of screenings of films from non-socialist countries increased dramatically: from 182,681 in 1960 to 322,323 six years later. There was a correspondingly rapid drop in the number of screenings of Soviet films from 240,982 in 1960 to 59,546 in 1967, and declining to 19,536 screenings over the next two years, in part as a result of the extraordinary political situation after the Soviet invasion in August 1968 (Figure 6.1). Despite the continuing decrease in total attendance, the number of viewers seeing films from non-socialist countries grew from 36,268,000 in 1960 to 49,525,000 in 1966 (Figure 6.2). In contrast to 1963, when the five-year economic plan collapsed, imports of films from capitalist countries increased significantly in the following year.[49] Despite the continuing lack of foreign currency, the ÚV KSČ Secretariat considered it appropriate to grant special foreign currency allocations for the purchase of films, which would bring 'quite outstanding revenues' when distributed, despite their high cost.[50] While only four US films were distributed in 1963, there were eight in 1964, including the hits *Big Country* (William Wyler, USA, 1958), *Some Like It Hot* (Billy Wilder, USA, 1959), *Roman Holiday* (William Wyler, USA, 1953), and the most popular one, *The Magnificent Seven* (John Sturges, USA, 1960).[51]

This emphasis on the economic performance of film distribution was regarded as negative after 1969, in the era of the so-called 'normalization' and repression of the 1960s reformist effort. An official report for the presidium of ÚV KSČ on the Czechoslovak film industry argued that 'the economic interest of distribution … leads to a general preference for commercial films, thus shifting the overall composition of films introduced by the distribution network to an area of entertaining, adventure, and otherwise escapist films'.[52] Nevertheless, before normalization affected the film industry with its purges beginning in September 1969 and before the distribution supply was 'cleansed' by excluding 32 'Western' films in 1971, the Czechoslovak Film Company achieved the highest receipts in its history in 1969; in addition, the number of screenings of films in the 'other production' category was the greatest for the entire post-war period.[53]

In Figures 6.1 and 6.2 we can see the massive fluctuation in the attendance rate and the number of screenings of Western films on the one hand and Soviet and 'popular democratic' ones on the other. These fluctuations were in part caused by the responses of both audiences and the distribution network to the invasion of the intervention armies in August 1968. On 18 December 1968 the Government demanded inclusion of 'artistically valuable films from the Five Countries' in the programming.[54] The local distribution agencies were able to find excuses for some time by saying that despite their willingness to cooperate, economic necessity forced them to programme Western films because they yielded higher receipts. Nonetheless, the programming department had already issued an order to schedule one Soviet or 'popular democratic' film per month regardless of the cinema managers' preferences.[55] This was the final example of the persistent contradiction between the ideological and the economic requirements in the period in question, and in the era of so-called 'normalization' after 1970, the contradiction was firmly resolved in favour of the ideological imperative by the political elite.

Conclusion

In this chapter I have approached the history of cinema distribution in a totalitarian regime primarily by looking at hierarchical power relationships and by using statistical data and archival material. The overview illuminates two features of the relationship between distribution practices and the ideological premises of the political regime. By studying the operation of the distribution system, we can examine why the regime's attempt to establish a new film culture failed. The inability to change audiences' cultural preferences to those that were officially desirable undermined the economic foundation of the cinema network by restricting box-office revenues. The inefficiency of the socialist economic system limited the amount of support provided by public revenue.[56] The conflict of interests between different institutions, political authorities and economic subjects in the system, and their competing goals, ensured that distribution practice veered from one position to another, neither fulfilling its ideological purpose nor achieving its economic objectives.

The regime also saw cinema distribution as a tool for implementing the socialist version of modernization. The ideological discourse embedded the cinema – both the movies themselves and their supposedly democratic, prompt, omnipresent distribution – in the modernizing project of enlightenment and education. Distribution practices were subordinated to the duty to educate a new kind of viewer (and in the wider frame of the cultural revolution, a 'new kind of man'), and with it the dissemination of new socialist values and the 'cleansing' of cultural space. The system of distribution was, however, itself a symbol of the new order, and this role resulted in ideological goals taking precedence over economic ones, at least until 1960. In the 1960s, the system slowly shifted from one rationalized according to ideological values towards one rationalized according to economic values. Because the cinema's infrastructure and technology was an emblem of industrial modernization, the development of the cinema network and shifts in technical and aesthetic standards such as the adoption of colour and widescreen were all subordinated to the state plan, which was itself a response on the part of the proponents of state socialism to the challenge of modernization.[57]

In the first five-year plan, the pace and form of 'cinefication' as an infrastructural precondition for entry of the socialist values was explicitly determined by the Ministry of Culture and Education, as was the spread of widescreen cinemas.[58] But the ratio between movies from the capitalist and socialist countries and the number of their respective viewers was specified and planned only after the intended 'natural' shift of demand towards socialist production had failed. What distinguished the 1960s from the 1950s was the shift from uniformity and standardization to a greater diversification, a movement from asceticism to an acceptance of limited consumerism.[59] Instead of the prefabricated cinemas and haphazard increase of seats of the 1950s, the 1960s saw small cinemas close as larger ones modernized by installing widescreen projection. Travelling cinemas no longer spread the 'globalized' aesthetic model of socialist realism to the smallest villages. Instead, they were used to screen popular widescreen pictures, such as the Karl May film series, at holiday centres. In the early 1950s, summer cinemas showed the most ideologically charged films, but in the following decade they screened the most attractive 'summer hits' like Louis de Funès' comedies, French and Italian swashbucklers, or American pictures of whatever genre. This movement from an industrial modernization marked by austerity and standardization in the sphere of consumption to a period in which the state apparatus paid more attention to consumer goods and leisure activities brought with it more film images of the attractions of Western modernization.

To comprehend the socialist version of modernization, the historian has first to conceptualize the distinction between the Western and the socialist model of modernization, understand what is 'modern' in socialist societies that contain many anti-modern features, and evaluate the significant shifts in the socialist regimes' models of modernization throughout the decades of their existence.[60] At least until the early 1960s, indigenous film production was strictly controlled to promote socialist modernity. But the space of cinematic experience was also constituted by the movies imported from the West, and these movies were the most widespread and available consumer product from the countries behind the Iron Curtain. While the selection of Western pictures was tightly controlled and favoured socially critical movies and historical topics, the style, stars, technology, and the images of contemporary Western society that they made available contrasted with the promoted version of socialist values in ways that the regime's contradictory approach to cinema distribution could not entirely appropriate.

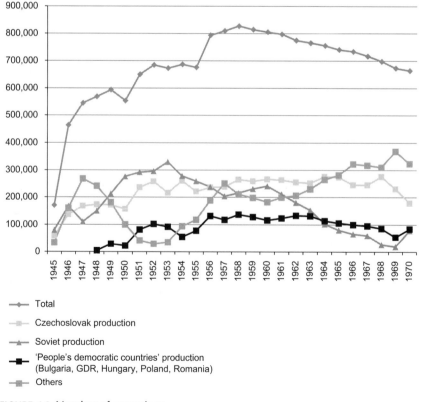

FIGURE 6.1 Number of screenings

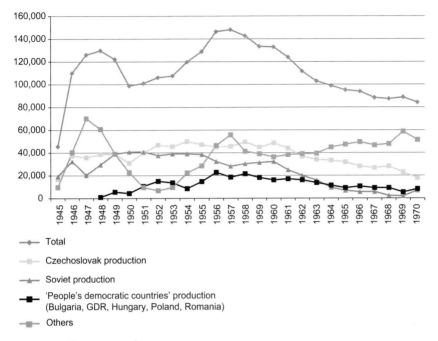

- ◆ Total
- ▪ Czechoslovak production
- ▲ Soviet production
- ■ 'People's democratic countries' production
  (Bulgaria, GDR, Hungary, Poland, Romania)
- ▪ Others

FIGURE 6.2 Cinema attendances

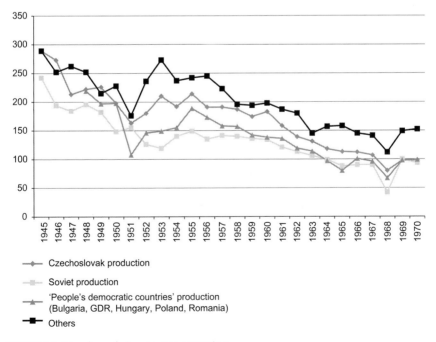

- ◆ Czechoslovak production
- ▪ Soviet production
- ▲ 'People's democratic countries' production
  (Bulgaria, GDR, Hungary, Poland, Romania)
- ■ Others

FIGURE 6.3 Number of viewers per screening

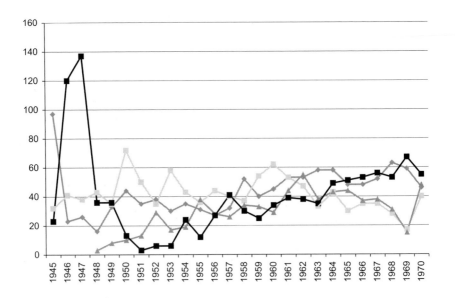

—◆— Czechoslovak production

—■— Soviet production

—▲— 'People's democratic countries' production
(Bulgaria, GDR, Hungary, Poland, Romania)

—■— Capitalist countries' production (USA, France,
Italy, Great Britain, West Germany)

FIGURE 6.4 Number of film programmes

Notes

This research was supported by Czech Science Foundation (research project no. 408/07/
P174) and by the Fund of the Faculty of Arts' dean, Masaryk University.

1 Although the 'Czech lands' Bohemia and Moravia were part of Czechoslovakia, this
chapter does not examine distribution in Slovakia.

2 'Ideová výchova a kulturní politika strany. Z referátu Václava Kopeckého na VIII. sjezdu
KSČ' (March 1946), in Z. Štábla and P. Taussig, *KSČ a československá kinematografie (výbor
dokumentů z let 1945–1980)*, Prague: Československý filmový ústav, 1981, p. 28.

3 The process was not limited to the cinema. The sphere of literature is analysed by
Pavel Janáček in *Literární brak. Operace vyloučení, operace nahrazení, 1938–1951*, Brno:
Host, 2004.

4 See P. Skopal, 'Filmy z nouze. Způsoby rámcování filmových projekcí a divácké
zkušenosti v období stalinismu', *Iluminace* 21, 2009, 71–92. For the concept of the 'New
Socialist Man' of which the model of a new cinemagoer was a specific derivative, see
D.W. Paul, *The Cultural Limits of Revolutionary Politics. Change and Continuity in Socialist
Czechoslovakia*, New York: Columbia University Press, 1979, pp. 16–52; E. Taborsky,
Communism in Czechoslovakia 1948–1960, Princeton: Princeton University Press, 1961,
pp. 469–607.

5 This leftist opinion of the Czech cultural elite was manifested before the first post-war
parliamentary elections, when over 800 of them signed a proclamation supporting

communist ideas. See A. Kusák, *Kultura a politika v Československu 1945–1956*, Prague: Torst, 1998.

6 The first proposals for the nationalization of the Czechoslovak film industry were prepared in 1942. Of the Soviet Bloc countries, the film industry was nationalized in 1945 in Poland, in 1947 in Albania and Yugoslavia, in 1948 in Bulgaria, Romania and Hungary. See G. Vincendeau (ed.) *Encyclopedia of European Cinema*, London: Cassell/BFI, 1995. The development in the Eastern part of Germany was complicated; the feature film studio DEFA was nationalized in 1953. See D. Berghahn, *Hollywood Behind the Wall. The Cinema of East Germany*, Manchester: Manchester University Press, 2005, pp. 11–54; J. Feinstein, *The Triumph of the Ordinary. Depictions of Daily Life in the East German Cinema, 1949–1989*, Chapel Hill/London: University of North Carolina Press, 2002, pp. 19–44.

7 For an analysis of the role that the 'sharp course' declared by the Party's Central Committee in September 1948 had in the sphere of culture, see J. Knapík, *Únor a kultura. Sovětizace české kultury 1948–1950*, Prague: Libri, 2004, pp. 120–24.

8 K. Pence and P. Betts (eds) *Socialist Modern. East German Everyday Culture and Politics*, Ann Arbor: University of Michigan Press, 2008, pp. 11–15.

9 See T. Martin, 'Modernization or Neo-Traditionalism? Ascribed Nationality and Soviet Primordialism', in S. Fitzpatrick (ed.) *Stalinism: New Directions*, London/New York: Routledge, 2000, pp. 348–67.

10 S.N. Eisenstadt, 'Multiple Modernities', *Daedalus* 129, 2000, 1–29.

11 D.L. Hoffmann, 'European Modernity and Soviet Socialism', in D.L. Hoffmann and Y. Kotsonis (eds) *Russian Modernity. Politics, Knowledge, Practices*, New York: Palgrave Macmillan, 2000, p. 247.

12 In the case of Czechoslovakia, the regime presented itself with a distinctly anti-modern slant as an heir of two prominent traditions: the first was 'Hussitism', referring to the doctrines of the reformist and nationalistic movement initiated by John Huss in Bohemia about 1402; the second was 'Czech National Revival', a cultural movement reviving Czech national identity during the late eighteenth and the first half of the nineteenth century. Both movements were promoted by the Czechoslovak regime and injected into film production. For a reflection on the co-presence of modern and anti-modern elements in the socialist countries, see also J. Kocka, 'The GDR: A Special Kind of Modern Dictatorship', in K.H. Jarausch (ed.) *Dictatorship as Experience: Towards a Social-Cultural History of the GDR*, New York: Berghahn Books, 1999, p. 20.

13 See, especially, K.H. Jarausch, 'Care and Coercion: The GDR as Welfare Dictatorship', in Jarausch, *Dictatorship as Experience*, pp. 47–69.

14 See V. Kepley, Jr, '"Cinefication": Soviet Film Exhibition in the 1920s', *Film History* 6, 1994, 262–77.

15 In 1963 there were 81 seats per thousand citizens in the Czech lands. By comparison, in 1965, Great Britain had 37 seats per thousand, France 54, and Poland 23 seats. Source of the data: L. Pištora, 'Filmoví návštěvníci a kina na území české republiky. Od roku 1945 do současnosti', *Iluminace* 9, 1997, 63–106.

16 'Pětiletý plán – Čs. státní film', in: *Ústřední výbor KSČ 1945–1955*, Central department for cultural propaganda and ideology, 1945–89, RG 1261/2/18, AU 662, NA.

17 For the use of the concept of cultural revolution in Soviet history, and a critique of its simplistic rendition by historians, see M. David-Fox, 'What Is Cultural Revolution?', *The Russian Review* 58, 1999, 181–201. According to David-Fox, the evolution of the concept was often limited to an opposition between Lenin's definition as 'mass education and cultivation of civilized behaviour' on one side and the militant class-war definition after 1928 on the other. The influential writing of Sheila Fitzpatrick located the cultural revolution in the specific historical situation of the First Five-Year Plan 1928–32 in the USSR: the term 'described a political confrontation of "proletarian" Communists and the "bourgeois" intelligentsia'. S. Fitzpatrick, *The Cultural Front. Power and Culture in Revolutionary Russia*, Ithaca/London: Cornell University Press, 1992, p. 115.

18 V. Kopecký, 'Vedeni nepřemožitelným učením marxismu-leninismu vybudujeme socialismus v naší vlasti', in *Protokol IX. řádného sjezdu komunistické strany Československa: V. Praze 25.-29. května 1949*, Prague: ÚV KSČ, 1949, p. 366.

19 See D.W. Paul, *The Cultural Limits of Revolutionary Politics*, pp. 16–52; D. L. Hoffmann, *Stalinist Values. The Cultural Norms of Soviet Modernity, 1917–1941*, Ithaca/London: Cornell University Press, 2003, p. 7.

20 M. Huláková, *O kultuře a kulturní revoluci*, Prague: Nakladatelství politické literatury, 1963, pp. 266–67; see also proclaiming the role of the cultural revolution for an evolution of the socialist man in the theoretical revue of KSČ *Nová mysl* ('New Mind'). Anon., 'Úvodník: Socialistická kultura a socialistický člověk', *Nová mysl* 6, 1959, 562–70.

21 F. Kahuda, 'Zvýšit ideovou účinnost a výchovné působení filmu, všestranně zkvalitnit činnost kin', in *Celostátní konference o filmové distribuci*, Prague: Ústřední půjčovna filmů, 1958, p. 5.

22 Archive of Ministry of Interior, 318 – Hlavní správa tiskového dohledu Ministerstva vnitra, daily report no. 217, 13 February 1959.

23 Anon., 'Proč dosud nehrají česká kina? Problém nových kopií a problém starých titulků', *Filmová práce* 2, 1945, 1.

24 Sojuzintorgkino was the Soviet company for export and import of films.

25 See P. Mareš, 'Politika a "Pohyblivé obrázky". Spor o dovoz amerických filmů do Československa po druhé světové válce', *Iluminace* 13, 1994; P. Mareš, 'Všemi prostředky hájená kinematografie. Úvodem k edici dokumentu "Záznam na pamět"', *Iluminace* 6, 1991, 75–105.

26 The state company for film enterprise, replaced by Czechoslovakian State Film in March 1948.

27 *Československá filmová společnost II* (unprocessed files), National Film Archive, Prague.

28 A letter of Josef Jonáš, the chief of organizational secretariat of the Cultural and promotional division of ÚV KSČ, 14 March 1947, addressed to Gustav Bareš (the chief of the Cultural and promotional division of ÚV KSČ). KSČ – ÚV 1945–89, ÚV KSČ Cultural and promotional division files. Archival unit 660, National Archive, Prague (hereafter ÚV KSČ, OKP, AU 660, NA).

29 A quotation from the presentation of Málek at the conference of the Czechoslovakian State Film in November 1948. In *Zpráva o první celostátní podnikové konferenci Československého státního filmu*, Prague: Československý státní film, 1948.

30 'Za vysokou ideovou a uměleckou úroveň československého filmu. Usnesení předsednictva Ústředního výboru KSČ o tvůrčích úkolech československého filmu', *Rudé právo* 19 April 1950, p. 3. *Rudé právo* was the main Czechoslovakian daily newspaper of the Communist era.

31 The model of 'circular distribution' was finally denied as 'evolutionarily permanently unbearable', and in 1953 only two programme sequences were put together. See J. Havelka, *Výroční zpráva Čs. státního filmu za rok 1952*, Prague: Ústřední ředitelství ČSF – Filmový ústav, 1964, p. 115.

32 ÚV KSČ, OKP, RG 666, NA.

33 Bohuslav Hammer, the Brno regional director. Inscription of the cinema heads, 28 January 1949, Krajský filmový podnik (unprocessed files), G 604, Moravský zemský archiv Brno (hereafter KFP, G604, MZA).

34 Jaroslav Brož, 'Choď'me do kina bez předsudků', *Svět práce* 20, 1949, 10.

35 An inscription of the deliberation at the film subcommittee of the Regional National Committee's school division, Brno, 12 December 1950 (KFP, G604, MZA).

36 A. Černík, *Výroční zpráva o čs. filmovnictví. Rok 1949*, Prague: Československý státní film, 1952, p. 160.

37 Working conference, Brno, 28 November 1950 (KFP, G604, MZA).

38 Ministry of Culture, 1953–56, AU 867, 196/141, National Archive, Prague (hereafter MK, AU 867, NA).

39 Ibid.

40 Anon., 'Putování po SSSR', *Kino* 23, 1951, 542–43.

41 The most common and omnipresent tool of framing the feature film experience by images of precipitously developing society ensured, of course, the newsreels. For a detailed analysis of the practice of framing in the 1950s, see Skopal, 'Filmy z nouze', 70–91.

42 Governmental decree, 16 January 1957, no. 4/1957 about organization of the film business: all property of cinemas including real estates was conveyed to the National Committees. At the same time regional film enterprises were established.

43 J. Havelka, *Čs. filmové hospodářství 1956–1960*, Prague: Československý filmový ústav, 1973, p. 184.

44 A report for ÚV KSČ Secretariat, given by the fourth department of ÚV KSČ Secretariat: a report about the contemporary cultural political difficulties of the film distribution. Ústřední výbor KSČ 1945–89, Prague – secretariat 1954–62, RG 1261/0/14, file 130, AU 192, item 3, NA (hereafter ÚV KSČ, secretariat 1954–62, AU 192, NA).

45 Economic pressure to allocate funds from distribution for film production led to the restricted use of the rule since 1964. 'O co vlastně jde? O diváka – o peníze – o ... ?', *Divadelní a filmové noviny* 13, 1965–66, 1; minutes from a meeting of the Regional Film Enterprise in Brno, 18 February 1964, KFP, G604, MZA.

46 F. Kahuda, 'Zvýšit ideovou účinnost a výchovné působení filmu, všestranně zkvalitnit činnost kin', 6.

47 Memo of the Central Film Rental Office, KNV Olomouc (1857) 1949–60, item II, 2068/1491, Country Archive Opava, subsidiary Olomouc.

48 See e.g.: 'Hlavní úkoly kulturní revoluce. Z usnesení XI. sjezdu KSČ (červen 1958)', in Štábla and Taussig, *KSČ a československá kinematografie*, pp. 77–78. See also the report for the ideological commission of ÚV KSČ about the contemporary situation in feature film production, presented by a member of the commission and employee of ÚV KSČ Secretariat Zdeněk Urban on 12 November 1959. 'Edice materiálů Banská Bystrica 1959. Dokumenty ke kontextům I. festivalu československého filmu', *Iluminace* 56, 2004, 200–209.

49 See e.g. P. Sirůček a kolektiv, *Hospodářské dějiny a ekonomické teorie (vývoj – současnost – výhledy)*, Slaný, 2007, pp. 193–97; and K. Kaplan, *Kořeny československé reformy 1968. II. část: reforma trvale nemocné ekonomiky*, Brno: Dopněk, 2000, pp. 234–306.

50 The Czechoslovakian Film Company's policy of export and import in the period 1957–63. ÚV KSČ, secretariat 1962–66, RG 1261/0/15, file 26, AU 48, item 1, NA.

51 Which is about 60 per cent of the population. *Big Country* was screened by the Světozor cinema in Prague for 19 weeks.

52 A report for the presidium of ÚV KSČ about the situation in the Czechoslovak film industry: 'Zpráva o situaci v československé kinematografii, kterou podával Jan Fojtík předsednictvu ÚV KSČ 6. ledna 1970. Dokumenty z archivu ÚV KSČ', *Iluminace* 25, 1994, 167.

53 Ibid., 191–99.

54 I.e., the states that participated in the intervention in Czechoslovakia: the Soviet Union, East Germany, Hungary, Poland and Bulgaria.

55 DFCVSE Brno minutes, 22 May 1970, ibid.

56 See J. Kornai, *The Socialist System. The Political Economy of Communism*, Princeton: Princeton University Press, 1992, for the economic analysis of the system.

57 P.C. Caldwell, *Dictatorship, State Planning, and Social Theory in the German Democratic Republic*, Cambridge: Cambridge University Press, 2003, p. 7.

58 Zpráva o zřizování cinemascopických kin a o podpoře kinofikace. (Report about cinemascopic theatres and support of cinefication). Source: RG, 994, Ministry of School and Culture, Session of the ministe's board 1956–66, board no. 41, NA.

59 For an analysis of an analogical, even if slightly achronological shift in the German Democratic Republic, see Ina Merkel's chapter with the telling title 'Consumer Culture in the GDR, or How the Struggle for Antimodernity was Lost on the Battleground of

Consumer Culture', in S. Strasser, Ch. McGovern and M. Judt (eds) *Getting and Spending. European and American Consumer Societies in the Twentieth Century*, Cambridge: Cambridge University Press, 1998, pp. 281–99.

60 For a survey and periodization of the modernization's course in the post-1945 European socialist regimes, see I.T. Berend, *Central and Eastern Europe 1944–1993. Detour from the Periphery to the Periphery*, Cambridge: Cambridge University Press, 1996.

7

'THE MANAGEMENT COMMITTEE INTEND TO ACT AS USHERS'

Cinema operation and the South Wales Miners' Institutes in the 1950s and 1960s[1]

Stefan Moitra

The relationship between the labour movement and cinema is a topic often forgotten in historiography today. A few *topoi* have found their way into the mainstream of film history, such as revolutionary Russia's cinematic experiments or the activism of left-leaning filmmakers in Weimar Germany and 1930s France. A number of studies in this context have also focused on the levels of audiences and exhibition. In British cinema historiography, Stephen G. Jones and Bert Hogenkamp have produced important monographs in the 1980s which explored the forms of production, distribution and means of audience organization linked to the Labour and Communist parties as well as to the trade unions.[2] So far, however, the main focus has been the interwar period while the decades after 1945 have mainly been left aside.

Generally speaking, when we use the term labour movement cinema practice, two different layers of meaning are addressed. The first concerns the production and distribution of films promoting ideas and values of the socialist labour movement.[3] The early, widely distributed Soviet productions and Dudow's and Pabst's Weimar films can be subsumed under this category along with educational and promotional films by trade unions or co-operatives.[4] The second layer of meaning regards the perspective of exhibition more directly and looks at the local uses of film exhibition organized by party or trade union groups as a means of both entertainment and attraction directed at local working-class audiences. In other words, we might distinguish a top-down perspective – employing films as an educational and politicizing tool – from the grassroots perspective of specific local uses of film exhibition. In reality, these interlinking levels of film practice led to a variety of cinematic activities practised by labour movement organizations in different countries. These could take the form of itinerant film shows during the Weimar Republic, small-gauge trade union shows of educational and neorealist feature films in post-war West Germany, the Italian Communist Party's production of documentaries and organization of audiences in the 1950s, or even the establishment of a whole distribution and exhibition structure by the Swedish labour movement since the late 1930s.[5] While the latter example was probably the exception to the rule, such exhibition practices might be described as what Richard Maltby and Melvyn Stokes have termed 'other cinema' – the uses of films outside the theatrical context of the cinema industry.[6]

I want to take up the idea of the 'other cinema' in what follows, although my use of the term will slightly expand its borders with regard to the supposed alternative means of film exhibition. The exhibition practice considered in this chapter was not situated outside but within the field of the cinema industry. At the same time, its protagonists were not commercial entrepreneurs but members of elected voluntary committees, representatives of cultural centres in small or medium-sized mining towns in a major British industrial region, the South Wales coalfield. The Miners' Institutes, Memorial, Welfare and Workmen's Halls that I will focus on functioned as self-organized cultural and social centres at the heart of the mining communities. Cinema operation was but one part of the Institutes' offerings, albeit a most important one both in terms of gaining revenue and meeting patrons' demands. Initially, mainly after the First World War, film exhibition was a questionable activity because, like labour movements elsewhere, Welsh workers' activists followed educational ideals along the lines of a high cultural canon. 'I do not believe it was the ideal to compete with other cinemas when the Welfare Halls were built. [...] What the Halls were for was to supply a permanent place where other types of culture should be nurtured', wrote Bert L. Coombes in his memoirs in 1945.[7] Those other types of culture, such as libraries, opera, choir and theatre performances, were part and parcel of the Institutes' programmes from the start. But in operating cinemas the Miners' Institutes developed a unique professionalism compared to the non-theatrical exhibition run by other labour movement organizations in Britain and elsewhere. From a business point of view and also in the eyes of the audience, this was not an alternative to 'proper' cinema – it *was* cinema. Its financial gains, nonetheless, were solely used to benefit the maintenance of the Halls in their function as self-organized community centres. In an area that largely lacked the infrastructure of mass culture, the Miners' Institutes sometimes represented the only form of institutionalized leisure apart from the public house, churches and chapels or sports. This initial lack of commercial competition led to an integration of the labour movement's educational values with the provision of 'mere' entertainment.

Given the post-war consensus on the welfare state and the subsequent boom of the consumer society, labour activists in other western European countries by and large refrained from maintaining separate, class-based, cultural institutions after 1945. In contrast, as long as the mass cultural gap was not otherwise filled, the Miners' Institutes of South Wales continued their cinema operations along with other programmes of leisure and culture well into the 1960s, and sometimes even longer. This specificity was sustained until two different lines of development led to a re-adjustment more in line with other western countries. On the one hand, changing consumers' demands resulted in new cultural habits entailing the decline of cinema audiences. On the other hand, like other European coalfields facing the competition of new energies, South Wales experienced a major industrial transformation. As a consequence, older community ties in the mining towns eroded, resulting in a fading need for community centres like the Institutes.

In this chapter, I will focus on the time between the end of the Second World War, a boom period for the Institute cinemas, and their subsequent decline in the late 1960s. Starting with an overview describing the beginnings of the Institutes' cinema activities, I will investigate their function as sites of modernization in the post-war coalfield society. After discussing the Institutes' level of professionalism and their role as independent entrepreneurs in the realm of the cinema industry, I will address the apparent change of cultural patterns after the mid-1960s, when audience numbers declined and the continued existence of the Institutes themselves was threatened.

Recreation and education: from the early period to the post-war era

The history of the Miners' Institutes reached back slightly longer than their earliest film exhibitions, before the First World War. Before South Wales acquired a reputation for workers' radicalism, donations by wealthy benefactors and company owners were used to build libraries and reading rooms for miners and their families after the region's development as a coalfield in the last quarter of the nineteenth century.[8] After the establishment of the Miners' Welfare Fund in 1921 as a joint board of the mining industry and the South Wales Miners' Federation, the Institutes were provided with a stable financial foundation by means of a fixed levy per ton of coal output paid by the employers. By then, however, industrial conflict had transformed the miners' political outlook and self-confidence from liberal industrial cooperation to a class-based, more or less socialist self-understanding.[9] Some Miners' Institutes had already been taken over by elected workers' committees before the establishment of the Miners' Welfare Fund, and this model of self-reliance could now be executed all over the coalfield. Responsibility for all activities was delegated to elected management committees which might be assisted by several sub-committees. Several cinema committees employed a manager to secure professional business, depending on the size of the hall. Built as a Temperance Hall in 1861, the Tredegar Workmen's Hall was taken over by a workmen's committee in 1911. One of the largest and oldest Institutes, it had a library, billiards, theatre and opera performances by youth groups and drama societies and let its rooms to community groups ranging from the Communist Party to the local brass band. From 1933, the Institute's main hall was used as a cinema.[10]

Some newer Institutes quite consciously presented themselves as epitomes of modern life within a context that, economically, was more and more crisis-ridden during the interwar period. When the Cwmllynfell Miners' Welfare Memorial Hall and Institute was opened in 1934, the founders proudly extolled the modernist layout of their site, which not only consisted of 'a bowling green, three tennis courts and a children's playground', but a main and a smaller hall with

> the usual stage services and retiring rooms, also a projection room and film store above the gallery [...] The [adjoining] Institute comprises a billiard room, library, games room, refreshment room and a kitchen, all opening out to

broad roof gardens. […] The construction and arrangement of the entire building represents the most modern practice of multi-purpose hall design. Every part of the building, except the main roof, is of reinforced concrete and consequently fireproof. […] The ceiling is treated with […] sound-absorbing materials which renders the auditorium acoustically perfect. The royal blue stage curtain, trimmed with gold, harmonises with the seating, while the proscenium surround is treated with fibrous plaster, above which is a neon light sun-ray effect, symbolising the dawn of a new era for the miners of Cwmllynfell and their families.[11]

Given the increasing rate of unemployment in the mining industry during the Depression era, mass demonstrations and even hunger protest marches by miners' families, not only the neon sun-ray but the modernist character of the Institute as a whole might have signified the necessary 'dawn of a new era for the miners of Cwmllynfell' and elsewhere.[12] While unemployment grew, users' demands also increased as the unemployed had more free time at their disposal.[13]

Against this backdrop, the operation of cinemas by the Miners' Institutes during the interwar years varied in its functions. In keeping with the leftist outlook of the mining communities, films were meant to be used for politicizing and educating. On the other hand, cinema could fulfil a demand for modern entertainment, both in the form of exhibiting popular films, and as architectural signifiers of modernity. Finally, there was the practical consideration that exhibition could be 'a very profitable enterprise'.[14]

Notwithstanding fears of mass cultural compromises, the ideological link between the labour movement and the Institutes' cinema programmes was strong. As Bert Hogenkamp has shown, some Institutes screened Soviet avant-gardist films by the likes of Sergei Eisenstein and Vsevolod Pudovkin in the 1930s. During the Spanish Civil War efforts were made to show films addressing the Republican cause, at times in connection with collections after screenings in support of the Spanish anti-fascists. Other Soviet productions with a stance against fascist Germany were shown before and during the Second World War. Nonetheless, Hogenkamp suggests that the Institute cinemas' biggest failure was that they did not manage to cooperate more closely with the Workers' Film and Photo League, the Progressive Film Institute and Kino Films, all of which represented centres of distribution and production for socialist and communist film culture in Britain at the time.[15] Apparently, films with an explicit political impact were part of the mining communities' cinematic experience, but not to the extent that it fully met Bert Coombes' demand for 'other types of culture'. The main part of the programme was made up of popular entertainment, but even here, as Robert James has recently pointed out, the choice of films corresponded to the Institutes' initial cultural and political categories.[16] Drawing on the example of the Institute cinema in Cwmllynfell, James has shown that the superficial impression of a merely entertaining programme can be reconsidered. Generically, the most popular programmes were made up of dramas and comedies, followed by musical and historical films and by adventures and romances. The dramas and comedies in particular, such as the Marx Brothers or the *Old Mother Riley* series, frequently addressed or parodied social issues and problems of class. Even the historical and musical titles 'demanded some measure of cultural competence from the audience'.[17] James argues that the cinema management committee in Cwmllynfell consciously chose films that served a sense of class consciousness and could be related to the values that the Institutes followed in their cultural and political outlook. Combining James' and Hogenkamp's analyses, it seems that even if the overall shape of the Miners' Institute cinemas was not solely constituted of films with an outright political content, the cinema programmes up to 1945 by and large matched the cultural demands that committees and workers alike wanted to see fulfilled. They were satisfying a working-class community that was eager to be entertained, but also conscious of a sense of its class position and political stands.

After the end of the war, this class-based cultural activism corresponded to the political and social measures implemented by the new Labour government. For South Wales, in particular, the decisive change was the nationalization of coal mining and the establishment of the National Coal Board as a governmental entrepreneurial body. In 1952 the Miners' Welfare Fund was replaced by the newly constituted Coal Industry Social and Welfare Organisation (CISWO). Among other objects, CISWO was charged with providing 'healthy recreation' and promoting 'the integration of workers in the coal industry in the communities in which they live'.[18] To secure the implementation of these obligations by providing the planned social activities in local communities, the National Coal Board was to pay CISWO 'as necessary', instead of paying a levy depending on the output of coal. For the Miners' Institutes, this meant a substantial improvement. They still relied on the income from cinema, but the economic threats that were implied in the old system of financing were now avoided. In addition, with the financial support both from the Miners' Welfare Fund and later from CISWO, it was possible after the war to carry out the necessary rebuilding of the Institutes' cinema halls. Twenty years after the discourse of modernization so apparent at the opening ceremony in Cwmllynfell, the very sites of the Institute buildings were again linked to the hope for a newly emerging affluence in the coalfield communities.

Institute cinemas as sites of modernization

From the outset, many Institute cinemas had to cope with a range of problems regarding their overall appearance, interior design and even hygiene. The minutes of the Tredegar Institute, for instance, report on the acquisition of two cats to solve the issue of rats afflicting the cinema hall in August 1944, and again a year later.[19] In another case the main walls were leaking, enabling 'damp from outside to penetrate through' into the interior.[20] A regular problem was bad seating. In a 1946 survey on the conditions of Institute cinemas, 21 of 30 reported being in need of new seating or repairs.[21] If no appropriate seats were available, people had to stand in the back or sit on the steps.[22] Other challenges, also connected to the problem of space, were issues of crowding, noise and the admission of children. In 1944 the Tredegar cinema committee agreed to project a slide before performances urging the audience to keep silent during performances and threatening that, 'if necessary, the picture [would] be stopped and offenders be threatened with permanent suspension'.[23] The cinema committee debated several times about admission hours for children under ten, about 'babies in arms […], i.e. in shawls' and finally warned parents: 'Unless they exercise control over small children further restrictions will be imposed'.[24]

As a platform to discuss common concerns, a Miners' Welfare and Workmen's Halls Cinema Association (commonly known as the Welfare Cinema Association) had already been initiated by the Miners' Welfare Fund in 1938.[25] With the outbreak of war, however, the start of its operation had been suspended. With the return of peace, the Association functioned as a base for the interchange of ideas among the Institutes and also as a mouthpiece for their common interests to the outside world.[26] Forty-five cinemas were members of the Association in 1946, with two more joining somewhat later. According to the 1946 survey undertaken by the Association, the three largest member cinemas were the Nixon's Workmen's Hall and Institute in Mountain Ash, the Abercynon Workmen's Hall and Institute, each with more than 1,000 seats, and the Brynamman Public Hall and Institute with 964 seats. The smallest ones were the Bedlinog Hall and Institute with 320, the Penclawdd Hall and Institute with 350 and the Cwm and Llantwit Institute with 420 seats. Most of the Institute cinemas held between 600 and 800 people.[27] A main objective of the survey was to gather enough information to negotiate with the supplying companies for reduced prices for bulk purchases. By 1950 even G.B. Kalee, the leading general cinema supplier and part of the Rank Organisation, had agreed concessions for members of the Welfare Cinema Association, if only on the basis of contracts with individual cinemas.[28]

After basic problems like seating had been solved, and the general technical supply was provided for, the Institute cinemas could go on to approach the broader issue of modernization more positively. Following the example of the metropolitan cinema palaces and reports in trade journals like *Kinematograph Weekly* on the latest standards in technology, customer service or catering in the 'ideal kinema', the Institutes tried to comply with the latest models of consumer culture for their audiences' experience.[29] For patrons of the Tredegar Institute cinema, for instance, the consumption of ice cream had already been a regular feature of the early 1940s. In 1949 the committee decided to present a more diversified choice, splitting the weekly order between two manufacturers in order 'to test the patrons' reactions'.[30] For the period between 1945 and 1961 the Tredegar accounts list as many as thirteen different manufacturers who only supplied ice cream or other sweets.[31] This was complemented by a newly integrated buffet including an 'electric geyser' for hot products. A buffet supervisor was specially employed to serve food, sweets, ice cream, nuts, drinks and cigarettes from afternoon until the last evening performance.[32] Other committees negotiated with manufacturers to let them operate ice cream cabinets in their premises.[33]

Features like these directly addressed audiences as individual consumers. Other modernizing steps were taken at the technical level. While the fitting of neon lights inside and outside buildings enhanced the modern outward appearance, the most important and complicated renewal was undertaken with regard to film projection itself. As soon as the industry was talking about widescreen formats, the Miners' Institutes were eager to obtain this new technology for their audiences. As they had little knowledge of the advantages and disadvantages of different widescreen options, the Welfare Cinema Association took the lead in getting information. At a conference in April 1954, members were urged not to enter 'into any arrangements for the installation of alternative methods of projection until they received the report of the Executive Committee which, it will then be appreciated, is of the utmost importance to the organization and its members'.[34] In Tredegar, proposals to install CinemaScope had already been made a year earlier in April 1953, when not only Scope but 3-D was considered. In the event, it took the Tredegar committee almost two years to decide on the installation of CinemaScope. Negotiations and enquiries with a range of firms were made. A delegation was sent to the Miners' Institutes in Caerphilly and Maerdy which had already installed CinemaScope facilities, while the manager was sent to a trade show of *White Christmas* (Michael Curtiz, US, 1954), the first feature film in the Vistavision format, to report on its differences.[35] After the decision to install CinemaScope was finally taken in November 1954, it took yet another ten months for the first films in the new format to be played, because of disagreements with the distributor, Twentieth Century Fox, on the terms of booking.[36] Even the smaller halls opted for expensive widescreen facilities. An article in *Kine Weekly* from 1957 highlighted the restoration of the Llanbradach Workmen's Hall and Institute, where the cinema had been severely damaged by fire. A new balcony was installed 'with the object to improve sight lines' towards the 'CinemaScope size picture' that was now 'visible from every seat'.[37] In what the trade paper described as 'a striking example of the ways these Workmen's Halls committees watch for trends', a similar-sized cinema in 'a Workmen's Hall at Ton Pentre, down in the Glamorgan mining valleys' had by then had widescreen equipment installed for three years.[38]

Modern features like the latest sound and image technology and a diversified food programme enhanced the appeal of cinema in South Wales in the 1950s just as elsewhere. While the profit motive underwrote the Institutes' readiness to modernize their cinema halls, there is also a different strand of meaning to the implementation of these attractions. The 'affluent society' was not particularly visible in the Welsh mining communities of the 1950s.[39] The unemployment of the immediate post-war years had just begun to decrease and wages were not high enough to allow for the enjoyment of a wide variety of consumer goods.[40] Affluence started to become visible in small acts of consumption, including the cinema. The modernized Institute cinemas conferred something of the glamour that was difficult to participate in elsewhere. More importantly, at the centre of the mining communities all over South Wales, they indicated the shape of things to come. From the perspective of the Miners' Institutes, however, this promise of individual prosperity was not enough in itself. Recreation was still linked to an educational objective, or at least to the obligation to serve the community as effective cultural centres. The next two parts of this chapter will focus on the Institute cinemas as exhibitors and their attempts to act as independent entrepreneurs in the cinema industry at the same time that they tried to hold on to their older educational commitment to their community.

Institute cinemas as entrepreneurs and exhibitors

As we have seen, the Miners' Institute cinemas succeeded to a large extent in solving the pressing material problems of the war and immediate post-war era. As well as modernizing their appearance, the Institute committees also had to engage in the complex system of exchange characteristic of the cinema industry in order to present their audience with an equally satisfying on-screen programme. While insisting on the status of being charitable non-profit organizations, they also had to come to terms with the business principles upon which the rest of the cinema trade was based. In this respect, the miners' cinemas occupied a fragile position, one they constantly had to defend in relation to other parts of the industry.

A first step towards acting corporately was the formation of the Welfare Cinema Association. Although it was not equivalent to other commercial lobby groups, it nevertheless acted as a forum through which the cinema committees could discuss common problems; and it could also act on their behalf. It could not, however, act as an official representative for the Institute cinemas or negotiate on their behalf with distributors without endangering the charitable status of the individual Institutes.[41] Since this enabled them to claim exemption from the entertainment tax and other levies paid by the rest of the exhibition sector, it was crucial that the Association would not be regarded by government as a trading association.[42]

One consequence of this exemption illustrates the limits of the Welfare Cinema Association's abilities. The film distributors' contracts with cinemas claimed a fixed percentage of their gross takings. As the gross takings of the Institute cinemas had increased as a result of their tax exemption, the renters were legally entitled to demand their share from what the Institute committees claimed to be indisputably theirs. Representatives of individual Institutes and the Association put their case to local Members of Parliament and the Treasury, and also lobbied a variety of influential figures, like J. Arthur Rank and the Minister of Health, Aneurin Bevan, who had grown up in Tredegar, but to no avail. At a meeting with the Kine Renters Society (KRS) in 1948, Institute representatives pointed out that 'ten thousand miners in South Wales certified as suffering from coal dust disease were signing on at the Labour Exchange and spending the remainder of their time at the Welfare Institute'. The 'full benefit of the monies derived from exemption' was therefore needed for the 'building and maintaining [of] Institutes, etc., and the provision of amenities which would enable those unfortunate men, the miners and their dependents to get some joy out of life'. The distributors could not apparently 'forget about the big business interest concerned', and 'no concession whatsoever' was received.[43]

Since the Association could not represent the Institutes in the capacity of a trade organization, the cinema committees were compelled to join commercial cinemas in the Cinema Exhibitors Association (CEA) as the appropriate collective body. Although some feared that the Miners' Institutes would not get 'consideration as a small cinema', because 'the interests of the CEA were only in larger concerns', the idea was to use the trade organization to lobby for their own interests 'in the fight against the KRS'.[44] Accordingly, the Institute cinemas tried to participate quite actively in the CEA. In the reports of the CEA South Wales and Monmouthshire Branch it is evident that many Institute representatives took part in debates and tried to encourage the CEA to support their stand. William Berriman, manager of the Workmen's Hall cinema in Hopkinstown, Pontypridd and a leading member of the executive committee of the Welfare Cinema Association, served for many years as a member of the CEA General Council. As a pressure group within the CEA, Institute representatives articulated concerns that were shared by other small and independent exhibitors. A case in point was the question of three-day booking. While the film distributors wanted exhibitors to show films for a longer period to make higher gains, smaller cinemas feared that they would not have sufficient audience numbers for more than three days. In 1956 a resolution protesting against the distributors' refusal to hire films for this length of time complained that it often took 'periods of twelve months or more' before the small exhibitors were able to get hold of certain films. As Berriman pointed out at a CEA-meeting, this 'was unfair to the small situations throughout the country … By that time the glamour is gone completely off those big films … and you find that the winner of today becomes a very ordinary picture'.[45]

In issues like these, the representatives of the Miners' Institutes managed to make their voice heard at the industry's corporate level. This was one way of having an impact on the terms of trade. A different level was in the day-to-day business between cinemas and film distributors. If the Institutes were firm in their positions, negotiations could be at least partially successful. The Tredegar Institute could avoid a six-day booking of *The Robe* (Henry Koster, US, 1953), the first spectacular Cinema Scope success, by agreeing to have extra matinees on the days of screening.[46] But the hard bargains with the distributors continued to be difficult, and the cinema committees had to remain active on all possible levels. More than once the Welfare Cinema Association had to be used as an unofficial organ in order to put pressure on the distributors. As early as 1948, discussions with 'a number of film renters … including General Film Distributors, Warner Bros., Metro Goldwyn Mayer, Fox Bros. and British Lion' brought an agreement on a maximum renters' share of 40 per cent of the cinemas' gross takings. At the same time, Paramount continued to claim 50 per cent. It was agreed 'that the information be passed verbally to managers of welfare cinemas […] suggesting at the same time a boycott on Paramount until they fall into line'.[47] In 1956 the Association even threatened their members with 'immediate expulsion' if they deviated from insisting on the 40 per cent ceiling, since some distributors 'intend[ed] to approach members for a higher maximum' than the established charge. 'It is hoped that you will continue to give your loyalty to the Association as you have done in the past and adhere to conference resolutions which are made in the best interests of all concerned.'[48] Such individual and corporate negotiating was the backbone of the Institute cinemas' practical functioning. Already in 1948, an executive member of the Association pointed out that 'had this organization not been in existence, we still would have been paying the whole 100% [entertainment] tax, and some of the Welfare Halls would have been on the borders of bankruptcy'.[49] Despite the financial stabilization after the nationalization of the mining industry, this characterization remained applicable in the following years. The

Institutes continued to depend largely on the income derived from cinema operations and consequently endeavoured to drive hard bargains with the renters.

As far as audience numbers were concerned, the Institute cinemas did not have to worry until well into the 1950s. Although no comprehensive data exists on actual audience numbers, it seems from the minutes and reports that the Institutes' clientele were satisfied with the programmes their cinema committees offered. Reacting to the success that other Institutes had with cinema operation, an increasing number of Welfare Halls began to adopt film exhibition. In 1949 the Tredegar committee assisted the Institute in Pengam, 'discuss[ing] details about the cinema that they were taking over'. On a minor level, the Institute in the mining village of Abertysswg hired a 16mm projector in 1952 to start having film shows regularly, though not on a daily basis.[50] On the other hand, the cinema committees were careful not to take any risks. Early shows catering for shift workers were rejected because 'such a step would be uneconomic to arrange, having regard to the number involved'.[51] Only on special occasions, like the introduction of CinemaScope, did they schedule matinees aimed at shift workers. These shows were meant to promote and stabilize the regular programme rather than cater for a particular audience group's regular demand.[52]

Community cinemas in decline

All this administrative effort and economic achievement were meant to serve two distinct purposes. First, of course, cinema was in high demand by the working-class clientele in the area. A symbolic site of modern life, it offered the possibility of participating in attractions – on-screen and off-screen – that were otherwise not available. Second, cinema produced the revenue by which the Institutes' other activities could be subsidized. For the period between 1945 and 1955, the cultural committee in Tredegar sponsored an operatic society, a music circle, a choral society, a drama society, even a chrysanthemum society; it also provided rehearsal rooms for the Town Band, organized ballet performances, art exhibitions and balls.[53] The income generated by the cinema operation covered a large part of this expenditure. Apart from its financial contribution, however, film exhibition remained to be incorporated into the wider cultural and social objectives in accordance with the more traditional labour movement values promoted by the Institutes. While the larger part of the cinematic programme was made up of popular features, attempts were made in some places to offer alternative types of film to those provided by commercial cinema operators. Once a month from 1944 until at least 1955, the Tredegar Institute Film Society showed decidedly 'artful' and 'politically aware' titles, from Eisenstein's *Ivan the Terrible* (SU, 1944), Carl Theodor Dreyer's *Day of Wrath* (Denmark, 1943), Roberto Rossellini's *Paisà* (Italy, 1946) to documentaries like *Forgotten Village* (Herbert Kline, US, 1941) with a script by John Steinbeck and a score by Hanns Eisler, *Land of Promise* (UK, 1946), directed by Paul Rotha, or even *Spanish Fiesta* (US, 1942), a short by Jean Negulesco staging a ballet by Rimski-Korsakov.[54] In the late 1940s and early 1950s, Soviet documentaries and feature films also occasionally turned up on the programme.[55] Like cinema operation at large, these special programmes were sufficiently successful for other halls to ask the Tredegar Institute cinema manager for help in establishing their own film societies.[56]

Children were, of course, principal targets for the educational use of film. After many complaints about unruly children in the 1940s, attempts were made to establish special children's programmes in Tredegar. A first effort in 1943 to cooperate with local headmasters to show 'educational films' for school children was followed by the establishment of a proper children's club with a children's matinee every Saturday, taking on the model of the major circuits like Odeon or Gaumont: even the club song was taken from Odeon.[57] Unlike the major circuits, the educational task was extended outside the cinema hall by, for instance, taking the 650 members of the Children's Cinema Club for an excursion to the Bristol Zoo.[58]

All this can be seen as an integral part of an intact social setting in which the Miners' Institutes and their cinemas played a significant role. When this milieu began to change in the late 1950s, the Institutes as a whole had to struggle to stay alive. The mining communities were exposed to change on the industrial level, as more and more pits were closed as unprofitable on the international as well as on the domestic market. Between 1957 and 1959 alone 23 collieries in South Wales were closed. The process was acutely painful. The unity of work and leisure experience in neighbourhoods was destroyed. Men made redundant either sought new employment in collieries elsewhere or in newly established manufacturing industries nearby.[59] If the Miners' Institutes had been seen as centres of community life in their towns, they were now confronted with a basic alteration in their very clientele. An episode from Pontardulais illustrates this well. When, as late as in 1974, a representative of the National Union of Mineworkers organized a tour for a Hungarian choir and expected 'full support' from the local lodges, the chairman of the Institute committee pointed out that 'these days the local lodges did not comprise of only local men, there were a large majority who do not reside in our area and who would not be interested'.[60] By that time the communities had changed, and so had the Institutes.

The Institutes' cinema audiences declined along with the number of patrons for other activities. But a decline in cinema attendance had wider consequences for the Institutes, because of its effect on their income. Apart from the erosion of community structures, the cinema committees also had to come to terms with new interests and leisure practices. As early as 1953, the Tredegar committee decided to 'make a two months survey of the effect on takings when special programmes are on television'.[61] Yet, television and the other 'new gods' of consumer durables provided no big threat at this point.[62] The first 'intimation' of cinema closures was 'being received' by the Welfare Cinema Association in 1960, which recognized that 'this does not concern all our cinemas'.[63] The changes in the industrial and social makeup of the region were flanked by concomitant changes in patrons' demands. Starting in the early 1960s, gaming in general, and bingo in particular, became attractive as an alternative to cinema entertainment for adult audiences. By 1961 the Miners' Halls had to a large extent complied with these wishes and started to offer new gaming facilities within their premises. In February 1961, the Welfare Cinema Association circulated legal advice to be observed when games like 'housey housey' or tombola were played.[64] In June 1964, members were informed about special arrangements to show *Ben Hur* (William Wyler, US, 1959) that involved an exemption from a six-day booking where Institute cinemas had one or two regular nights of bingo.[65]

If the adoption of gaming mainly addressed an older audience, the extension of activities to licensed clubs and especially dance halls was directed at younger people. As well as moving away from the Institutes' educational aims, the new programmes also brought new problems. First, the legal position when moving beyond the traditional welfare activities was unclear. Taking this new challenge into account, the Association was reconstituted in 1965 under the name of Miners' Welfare Association with new separate sub-sections for cinemas, licensed clubs and one for cultural activities, the latter of which would 'give schemes without a cinema or a club the opportunity to be represented'.[66] The very term 'cultural activities' was now used as a catch-all for an assortment of programmes, its thin disguise hinting at a deeper change in the Institutes' self-image. The explicit educational scope that had been part of their constitution had largely vanished. The Institutes could not, however, compete with 'the full time and larger Bingo operators in the area' because of the 'very high stake money offered' by them.[67] In any case, the promotion of newer types of popular leisure activities could not replace the past financial success of cinema. Both cinema and bingo might have coexisted in the Welfare Halls in the 1960s, but as long as sufficient audience numbers remained, 'film performances were the main revenue'. As a report from the Pontardulais Welfare Hall pointed out, this was the case 'up until mid-1967', but in the summer of 1968 cinema in Pontardulais was reduced to one adult and one children's performance on Saturdays, while two nights were reserved for bingo. At the same time it was agreed to make all staff, except for the manager, redundant. To safeguard proper performances, the manager 'was given to understand that the management committee intend to act as ushers during the Saturday performances'.[68] Not much was left besides that. Several community groups still used the rooms for annual meetings, but other activities organized in the Institute amounted to little more than children's pantomimes and occasional jumble sales. After cinema receipts had dropped even further in 1970, the decision was finally taken to close cinema operation in Pontardulais, 'which in effect means that the regular business of the Hall will cease'.[69] Institutes and Halls such as this had struggled valiantly to retain their status as core centres in their communities, but by the end of the 1960s, public interest in them had virtually disappeared. At the annual general meeting of the Pontardulais Institute only 'eight persons from the general public' were present to discuss the desperate situation in the former heart of the community.[70]

Conclusion

All this might sound a bit too much like a narrative of decline starting with a 'good' form of working-class culture and ending with a 'shallow proletarian capitalism based on drinking clubs, bingo halls and cheap holidays abroad', as Welsh historian Kenneth O. Morgan has framed it.[71] My emphasis here lies on the activism with which the Institute committees maintained cinema operation well into the 1960s. In the first half of the century, the Institutes started out with a self-image as part of the international socialist labour movement, catering to all working-class needs including the provision of education and entertainment. The democratic principle of self-organizing was an obvious feature in this context and was to some extent transferred to the operation of cinemas, in the delegation of responsibilities to the managing committees, a labour movement cinema practice that continued well after the Second World War. The choice of films, however, was hardly linked to a wider plan of alternative leftist cultural politics. Rather, both in content and appearance, the managing committees tried to provide their communities with what was perceived as the latest standard of entertainment. As political or ideological questions with regard to programming played only a minor role, the Institutes' character moved closer to being a kind of consumer organization for local cinema audiences. On this level, the struggle for good renting conditions and tax exemptions became even more important as the purpose of the cinema operation was to enable 'the miners and their dependents … to get some joy out of life'. The committees certainly tried to get the most appealing pictures, although the insistence on three-day booking implies that audiences were probably more concerned with the act of going to the cinema than with the actual film. This principle of short-term booking was broken only very rarely, if a film promised to be really successful, such as *Gone With the Wind* (Victor Fleming, US, 1939).[72] Visiting the cinema was, of course, about seeing films, but it was also an act of participation in a form of cultural consumption that linked audiences, in the mining region just as elsewhere, with a sense of modernity, fashion and international stardom. The Institutes' constant attempts to modernize the outward appearance and projection technologies of their cinema halls underpinned this aspect. But for all that, this sense of modernity continued to be produced within the very working-class communities of which the Institutes were a part.

In their appearance and business structures, the Miners' Institute cinemas might not really seem to be examples of an 'other cinema' providing an alternative to conventional modes of exhibition. Their democratic, non-profit, grassroots structure of cinema operation made them exceptional, however. Profits were used only to make cinema operation self-sufficient and to finance other activities, many of which engaged the high cultural canon initially promoted in the Institutes. In this way, the cultural and political ideals of activists as well as the audiences' demands for entertainment were fulfilled. Only after those demands changed during the 1960s did the concept of self-organized Welfare Halls become outdated.

Notes

1 I would like to acknowledge the generous support for research on this chapter from the Glamorgan County History Trust. Thanks are also due to Elisabeth Bennett and Sue Thomas of the South Wales Coalfield Collection and Mr John Skinner who shared his knowledge on the matter. I want to dedicate this chapter to the memory of Nina Fishman.

2 B. Hogenkamp, *Deadly Parallels. Film and the Left in Britain 1929–1939*, London: Lawrence and Wishart, 1986; S.G. Jones, *The British Labour Movement and Film, 1918–1939*, London/New York: Routledge & Kegan Paul, 1987.

3 I deliberately use the singular form here for the sake of definition. One should be aware, nonetheless, of different national traditions and developments in the formation process of labour movements in different countries. Cf. K. Tenfelde, 'Europäische Arbeiterbewegung im 20. Jahrhundert', in D. Dowe (ed.) *Demokratischer Sozialismus in Europa seit dem Zweiten Weltkrieg*, Bonn: Historisches Forschungszentrum, FES, 2001, pp. 9–39.

4 For the British co-operative movement and its films see: A.G. Burton, *The British Consumer Co-operative Movement and Film: 1890s-1960s*, Manchester: Manchester University Press, 2005.

5 D. Langewiesche, 'Das neue Massenmedium Film und die deutsche Arbeiterbewegung in der Weimarer Republik', in J. Kocka *et al.* (eds.) *Von der Arbeiterbewegung zum modernen Sozialstaat*, Munich: Oldenbourg, 1994, pp. 249–64; S. Moitra, '"Reality is there, but it's manipulated": West German Trade Unions and Film after 1945', in V. Hediger and P. Vonderau (eds) *Films that Work: Industrial Film and the Productivity of Media*, Amsterdam: Amsterdam University Press, 2009, pp. 329–45; S. Gundle, *Between Hollywood and Moscow. The Italian Communists and the Challenge of Mass Culture, 1943–1991*, Durham, NC and London: Duke University Press, 2000; M. Jönsson and P. Snickars (eds) *Medier och politik. Om arbetarrörelsens mediestrategier under 1900-talet*, Stockholm: Statens ljud- och bildarkiv, 2007.

6 R. Maltby and M. Stokes, 'Introduction', in R. Maltby, M. Stokes and R.C. Allen (eds) *Going to the Movies. Hollywood and the Social Experience of Cinema*, Exeter: Exeter University Press, 2008, p. 9.

7 B.L. Coombes, *Miners Day* [1945], cited in P. Miskell, *A Social History of Cinema in Wales, 1918–1951. Pulpits, Coal Pits and Fleapits*, Aberystwyth: University of Wales Press, 2006, p. 150.

8 J. Rose, *The Intellectual Life of the British Working Classes*, New Haven, CT and London: Yale University Press, 2002, pp. 237–55; Ch. M. Baggs, 'The Miners' Institute Libraries of South Wales 1875–1939', in H. Jones and E. Rees (eds) *A Nation and its Books. A History of the Book in Wales*, Aberystwyth: National Library of Wales, 1998, pp. 297–305.

9 Cf. as the definitive work on the social and political history of the coalfield: H. Francis and D. Smith, *The Fed. A History of the South Wales Miners in the Twentieth Century* (2nd edition), Cardiff: University of Wales Press, 1998.

10 D.J. Davies, *Ninety Years of Endeavour. The Tredegar Workmen's Hall 1861–1951*, Cardiff: Tredegar Workmen's Institute Society, 1951.

11 Cwmllynfell Miners' Welfare – Memorial Hall and Institute. Programme of Opening Ceremony and Entertainment Week, 8–15 September 1934. South Wales Coalfield Collection (hereafter SWCC), no reference number.

12 K.O. Morgan, *Rebirth of a Nation. Wales 1880–1980*, Oxford: OUP, 1982, pp. 210–40.

13 Rose, *Intellectual Life*, p. 251.

14 Cf. the title quote of Robert James, '"A Very Profitable Enterprise": South Wales Miners' Institute Cinemas in the 1930s', *Historical Journal of Film, Radio and Television* 27, 2007, 27–61.

15 Hogenkamp, *Deadly Parallels*, pp. 140–45; Hogenkamp, 'Miners' Cinemas in South Wales in the 1920s and 1930s', *Llafur* 4, 1985, 64–75.

16 James, '"A Very Profitable Enterprise"'; S. Ridgwell, 'Pictures and Proletarians: South Wales Miners' Cinemas in the 1930s', *Llafur* 7, 1997, 69–80.

17 James, '"A Very Profitable Enterprise"', p. 42.

18 *1st Annual Report of the Coal Industry Social and Welfare Organisation*, London: CISWO, 1952, p. 13.

19 Tredegar Workmen's Institute, Cinema Committee Minutes, 1940–56, SWCC MNA/NUM/I/38/6 (hereafter Tredegar Cinema Committee), 2 August 1944, 6 September 1944, 27 February 1944.

20 In this case the cinema was not run by a committee, but the Institute let the cinema to a commercial operator. Abergorki Workmen's Hall and Institute, Secretary's Correspondence, SWCC MNB/NUM/I/3/C28–32, Correspondence with Globe Cinema, Cardiff, 4 May 1945.

21 Miners Welfare and Workmen's Halls Cinema Association, Cinema Executive Committee Minutes, 1938–50, SWCC no ref. no. [hereafter Welfare Cinema Association, CEC], 5 April 1946.

22 Tredegar Cinema Committee, 31 March 1943, 26 May 1943, 30 June 1943.

23 Ibid., 18 May 1944

24 Ibid., 24 September 1943. Other examples: 28 May 1942, 4 March 1943, 1 September 1943.

25 Although the Miners' Welfare Fund was meant to cater for the needs of all British coalfields, the Welfare Cinema Association did only concern the Miners' Institutes as a specific feature of the South Wales coalfield.

26 See correspondence in Welfare Cinema Association, CEC, 1938–50.

27 Welfare Cinema Association, CEC, 5 April 1946.

28 See Welfare Cinema Association, CEC, 3 November 1949, 1 March 1950.

29 'The Ideal Kinema' was the title of a *Kinematograph Weekly* supplement.

30 Tredegar Cinema Committee, 2 March 1949.

31 Tredegar Workmen's Institute, Cinema Account Book, 1945–61, SWCC MNA/NUM/I/38/18 [hereafter Tredegar Cinema Account Book].

32 Ibid. and Tredegar Cinema Committee, 16 March 1949.

33 Pontardulais District Memorial and Welfare Hall, Minutes, 1964–75, West Glamorgan Archive Service D/D/PMWH.1 [hereafter Pontardulais Welfare Hall], 14 September 1971.

34 Welfare Cinema Association, Circulars 1951–68, SWCC no ref. no., 3 April 1954.

35 Tredegar Cinema Committee, 15 April 1953, 21 July 1954, 29 September 1954, 6 October 1954.

36 Ibid., 17 November 1954, 8 December 1954, 23 March 1955, 4 May 1955, 14 September 1955.

37 'Welsh cinema is rebuilt,' *Kinematograph Weekly*, 7 November 1957, p. 39.

38 'And Scope goes into Village,' ibid., 27 May 1954, 4; 7.

39 See among others: L. Black and H. Pemberton (eds) *An Affluent Society? Britain's Post-War Golden Age Revisited*, Aldershot: Ashgate, 2004.

40 Morgan, *Rebirth of a Nation*, pp. 317ff., 345ff.

41 Welfare Cinema Association, CEC, 5 April 1946.

42 Ibid., 2 August 1946.

43 Ibid., 27 February 1948.

44 Ibid., 24 November 1948.

45 'Protest against hold-up on three-day bookings', *Kinematograph Weekly*, 5 January 1956, p. 9.

46 Tredegar Cinema Committee, 23 March 1955, 6 April 1955, 27 April 1955, 4 May 1955.

47 Welfare Cinema Association, CEC, 6 August 1948.

48 Welfare Cinema Association, Circulars 1951–68, 5 December 1956.

49 Welfare Cinema Association, CEC, 24 November 1948.

50 Tredegar Cinema Committee, 2 February 1949, 6. August 1952, 19 November 1952.

51 Ibid., 18 June 1952.

52 Tredegar Cinema Committee, 23 March 1955; 'Welsh cinema is rebuilt', *Kinematograph Weekly*, 7 November 1957, p. 39 for Llanbradach.

53 Tredegar Workmen's Institute, Cultural Committee Minutes, SWCC MNA/NUM/I/38/9.

54 Tredegar Cinema Account Book, 1945–61, 9; 34; 109.

55 See Tredegar Cinema Account Book, p. 33; Report of a Conference of Representatives of the Miners' Welfare and Workmen's Halls Cinema Association, 5 March 1952, SWCC, Gwaun-Cae-Gurwen Welfare Association, MNA/NUM//I/18/C5.

56 Tredegar Cinema Committee, 7 March 1951. The Institutes asking for help were the ones in Blaina and Caerphilly.

57 Tredegar Cinema Committee, 9 October 1946.

58 Ibid., 9 June 1954.

59 Morgan, *Rebirth of a Nation*, pp. 317, 322.

60 Pontardulais Welfare Hall, Minutes 1964–75, 8 September 1974, West Glamorgan Archive Service, D/D/PMWH.1.
61 Tredegar Cinema Committee, 27 May 1953.
62 D. Docherty, D. Morrison and M. Tracey, *The Last Picture Show? Britain's Changing Film Audiences*, London: British Film Institute, 1987, pp. 23–29.
63 Welfare Cinema Association, Circulars 1951–68, April 1960.
64 Ibid., 3 February 1961.
65 Ibid., 10 June 1964.
66 Ibid., 19 January 1965.
67 Pontardulais Welfare Hall, Minutes 1964–75, 11 April 1968.
68 Pontardulais Welfare Hall, Letters, Saunders (cinema manager) to Jones (CISWO), 19 June 1968, 14 January 1969, 28 October 1969.
69 Pontardulais Welfare Hall, Minutes 1964–75, 28 April 1970.
70 Ibid., 11 August 1970.
71 Morgan, *Rebirth of a Nation*, p. 319.
72 29 March to 4 April 1948. Cwmllynfell Miners and Welfare – Memorial Hall and Institute, Cinema Manager's Diaries, 1946–55, SWCC, no reference number.

Part II
Audiences, Modernity and Cultural Exchange

8

URBAN LEGEND

Early cinema, modernization and urbanization in Germany, 1895–1914

Annemone Ligensa[1]

> The psychology of the cinematograph's triumph is the psychology of the big city – not only because the big city is the natural focus of all emanations of social life, but particularly because the big city psyche, always rushed, reeling from one fleeting impression to another, curious and unfathomable, is precisely the psyche of the cinematograph!
>
> Hermann Kienzl[2]

Early cinema between tradition and modernization

It has become a commonplace that cinema, modernization and urbanization are connected, especially since the so-called 'spatial turn' in cultural studies.[3] However, within the 'modernity theory' that mainly focuses on films and discourses, the relationship between the three phenomena is often highly abstract, while empirical studies of cinema as a media institution usually either refrain from addressing larger theoretical issues or have a fairly narrow geographical scope.[4]

I would like to attempt a multi-level analysis of this relationship, by looking at discourses and institutional practices as well as at films, using Germany as a case study.[5] Germany is of special interest, because Berlin is often regarded as a paradigm of the relationship between cinema and modernization, since it experienced a particularly dynamic period of development around 1900, coinciding with the emergence of cinema.[6] Many theorists of modernization were German, including some of the key thinkers most frequently referred to in reflections on the relationship of cinema to modernity, such as Max Weber, Georg Simmel, Walter Benjamin and Siegfried Kracauer. *Berlin – Die Symphonie der Großstadt* (*Berlin – Symphony of a City*, 1927) is one of the paradigmatic city symphony films in which modern city life was presented in a modernist style.

In accord with critics of the so-called 'modernity thesis' that early cinema was essentially shaped by and reflects modern perception, I share the concern that concepts of modernity are often teleologically projected from a later period, and that arguments are usually based on a small canon of films.[7] To this I would add that despite taking culturalist positions, proponents of the modernity thesis give surprisingly little consideration to cultural specificity. The appearance of cinema, for instance, did not coincide with marked urbanization in all countries. Moreover, attitudes towards both modernization and cinema in early twentieth-century Germany were often extremely negative, much more so than in other western countries.[8] Despite these reservations, exploring cinema's relationship with urbanization seems justified and fruitful, because it was already a contemporary concern. In Germany, the *Kinodebatte* ('debate on cinema') was conducted as vehemently as the discourse on modernization. While the cinema and the big city were often regarded as closely related, attitudes toward this relationship varied considerably. A 1914 article in the *Allgemeine deutsche Lehrerzeitung* exemplified the optimistic viewpoint:

> The cinematograph can bring the different groups of our people, employers and workers, town and country, industrialists and craftsmen, closer together and promote harmony, if the right kinds of films are shown.[9]

By contrast, Albert Hellwig, writing in the *Archiv für Kriminalanthropologie und Kriminalistik*, expressed a negative view of modernity, cinema and city life, which was much more typical of German intellectuals at the time:

> Naturally, the [negative effects of the cinema] will find a better breeding ground in a metropolitan milieu than in a small town or the country. On the other hand, through literature and films the acquaintance with the metropolitan morass will be spread to the country, a further decidedly corrupting effect.[10]

Despite the undeniable novelty of cinema's specific aesthetic characteristics and its wide accessibility, many of the arguments and concerns in this debate had preceded cinema and had already been applied to other forms of entertainment, such as popular fiction, world fairs and variety theatre.[11] Hence it is necessary to go beyond discourse analysis in exploring the relationship between early cinema and urbanization. In this chapter modernity will be understood as referring to developments and phenomena that were regarded as new and culturally significant by contemporaries.[12] I will argue that, especially in the case of Germany, the turn from the nineteenth to the twentieth century is best characterized as a radical clash of old and new, which provoked both fascination and fear. Early cinema was 'suspended between tradition and modernity' and profoundly shaped by both.[13] Ben Singer has recently termed the paradoxical and ambivalent relationship of early cinema with modernization 'ambimodernity'.[14] My argument will be advanced in two parts: first, on the level of cinema as a media institution, and second, on the level of films. The empirical material for my analysis comes predominantly from the research project 'Industrialization of Perception' at the University of Siegen (2002–9) as well as the research project 'Visual Communities: Relationships of the Local, National and Global in Early Cinema' at the University of Cologne (2010–).[15] This material comprises:

1 a digital database of film supply in Germany between 1895 and 1920 (c. 45,000 films);

2 a digital database of film programmes of permanent cinemas between 1905 and 1914, compiled from the newspapers of nine German cities of different regions and sizes (c. 1,200, containing c. 3,800 different films);

3 a digital database on itinerant cinemas in Germany between 1896 and 1926 (with information on showmen, locations and events);

4 a digital collection of c. 4,750 contemporary texts from a wide variety of newspapers, magazines and journals;

5 a large paper copy collection of contemporary announcements and reviews of many of the films of the programme database;

6 a large digital collection of early films.

This unique, extensive and diverse body of material provides the most thorough and systematic source we have for understanding what was typical of German cinema between 1895 and 1914.[16]

Cinemas in the city

Arguments about the relationship between cinema, modernity and the city usually begin with the claim that cinema first appeared in cities. This refers, often only implicitly, to the esablishment of permanent cinemas, which only began to appear in significant numbers around 1905. Film projection first appeared almost simultaneously in several metropolises in the mid-1890s. Immediately afterwards, and well before the appearance of permanent cinemas, film shows appeared in surprisingly diverse and geographically widespread contexts. Films were not only shown in urban variety theatres, but also by itinerant showmen in various multi-purpose locations such as hotels and restaurants. Travelling cinemas appeared in circuses and set themselves up at festivals, markets and fairs. All these exhibition practices continued long after the boom in permanent venues. Although these presentations were not permanent, they were often connected with established cultural contexts and regular events, and travelling with the cinematograph was relatively easy, so diffusion was rapid. The very first mention of the new entertainment technology in the German showmen's trade press in 1896 already contained advice on what to take into account when travelling with the cinematograph to small towns.[17]

The Siegen database on travelling cinemas reveals how much this form of exhibition has been underestimated in our understanding of early cinema in Europe (see Figure 8.1).[18] In Germany, travelling cinemas apparently reached a larger audience than urban variety theatres, and as the graded ticket prices of film shows suggest, visitors to festivals, markets and fairs came from all social strata.[19] For many of the agrarian poor, however, any form of entertainment still seems to have been a rare luxury at the time, as Max Hölz's autobiography (1889–1933) attests:

Number of locations

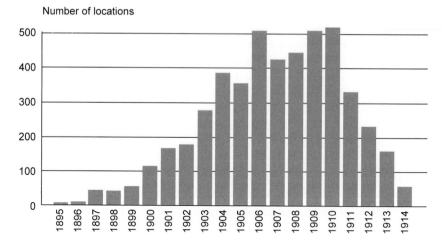

FIGURE 8.1 Diffusion of itinerant cinemas in Germany
Source: J. Garncarz, 'Über die Entstehung der Kinos in Deutschland 1896–1914', *KINtop: Jahrbuch zur Erforschung des frühen Films* 11, 2002, 153

> Until I was fourteen years old, I had taken part in only three children's amusements. The first was a school trip to the ruins of a monastery, the second was a puppet theatre show, ... and the third ... was a visit to a panopticon in the nearby town during a local festival.[20]

The fairground itself was an example of the simultaneity of tradition and modernity, as well as the reciprocal relationship and complex convergence of city and country. Markets were regular institutions in (pre-)modern cities, and in rural areas where they were temporarily set up they fulfilled urban functions such as trade, communication, socializing and entertainment. Based in old religious (especially Catholic) traditions, festivals became increasingly secularized and industrialized: travelling variety theatres were converted to film shows, rides became larger and more elaborate, and diffusion and accessibility were furthered by the railway system.[21]

By contrast, permanent cinemas (Figure 8.2) originated in the busiest areas of big cities, because as long as film supply was still relatively limited, a large audience base was needed for regular operation. One of the first types was called the *Ladenkino* (storefront cinema) in Germany, because these cinemas were often converted shops. Beyond the well-known comparison of cinemas and department stores based on the general similarities of display and commercialism, there was a more direct relationship: storefront cinemas were often founded by shop owners who could not or did not want to compete with department stores, which were spreading at the time.[22] Many storefront cinemas ran their programmes continuously, and in the daytime, walk-in audiences probably included shoppers. Interestingly, however, at least in Germany, department stores did not originate in big cities, but spread from provincial towns.[23]

Just as travelling cinemas were both integrated into established cultural contexts and contributed to their transformation, permanent cinemas and their urban environment shaped each other. Urban cinema represented a new type of 'cultural public sphere', in which diverse social groups participated and mingled to a new degree.[24] These heterogeneous 'urban masses' were a cause of concern for authoritarian Wilhelmine elites, and cinema was therefore subject to a wide variety of proposed and actual regulations, including censorship, licensing and taxation. At the same time, a process of social distinction through differentiation of exhibition venues was furthered by the film industry itself. Depending on the type of cinema and its location, different social spaces developed. Storefront cinemas were quickly followed by *Kino-theater* (cinema theatres) and after 1910 by *Kinopaläste* (cinema palaces), which targeted a sophisticated audience. *Kiez-Kinos*, cinemas located in Berlin workers' tenements, provided an intimacy that was almost like a village community within the metropolis.[25] In general, however, urban proletarians seem to have been the least likely to go to the cinema regularly, because most did not have the time or money for it.[26]

It was probably the rapid spread of permanent cinemas beyond big cities that literally pushed itinerant cinemas off the map.[27] Soon not even Simmel's 'metropolitan man' necessarily risked being bored when visiting a small town:

> Anyone who often travels alone knows the nearly endless, monotonous evenings one passes in small, dreamy towns, which attracted us with some old monument, a masterpiece of fine art or the commemoration of a historic event. At dusk, the last magic of such towns fades, and nothing remains but hopeless boredom, which one tries to combat without success. If one does not want to spend his entire time in smoke-filled, dreary inns or respectably sit out his stay in dining halls of small hotels decorated with old-fashioned plush and even worse oil prints, almost always the cinematograph beckons as the last refuge, to which today, both at home and abroad, places have aspired that have otherwise vehemently defended their virginity against culture.[28]

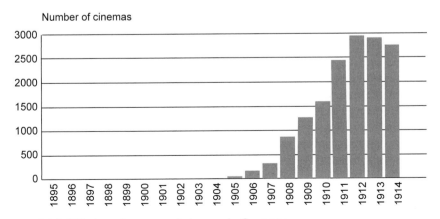

FIGURE 8.2 Diffusion of permanent cinemas in Germany
Source: J. Garncarz, 'Über die Entstehung der Kinos in Deutschland 1896–1914', p. 153

TABLE 8.1 Examples of cinema density in Germany in 1910

	Berlin	Hamburg	München	Leipzig	Würzburg
Number of cinemas	139	40	28	31	4
Inhabitants per cinema	14,285	23,250	20,000	40,000	21,000

Source: Absolute numbers of cinemas from A. Jason, *Der Film in Ziffern und Zahlen: die Statistik der Licht-spielhäuser in Deutschland 1895–1925*, Berlin: Deutsches Druck- und Verlagshaus, 1925. Population statistics from *Kürschners Staats-, Hof- und Kommunal-Handbuch des Reichs und der Einzelstaaten, nebst Anhang: Die außerdeutschen Staaten; zugleich Statistisches Jahrbuch*, München: Ertel, 1910.

Nevertheless, even in the 1920s the German landscape beyond towns of 10,000 inhabitants was described as a 'cinema desert' (*Kinowüste*).[29] Hence, another factor apart from diffusion must be considered: with the general increase of leisure time, money and mobility (bicycle use boomed around 1900), audiences could travel to the cinema with greater ease than the cinema could travel to its audiences. As early as 1913, contemporaries noted that 'especially on Sundays, large crowds of people travel from the country to neighbouring towns to attend cinemas'.[30] Still, at the time, weekly attendance in Germany was estimated at only 10 million.[31]

Despite the development towards a modern nation state, significant regional differences existed in Germany, but there has so far been little systematic study on how this influenced cinema.[32] Table 8.1 indicates that the number of cinemas in a city was not simply dependent on its size, suggesting that regional differences in censorship practices, reflecting varying attitudes towards cinema entertainment, may have significantly influenced cinema diffusion.[33] The debate over how the regulation of cinema was to be distributed between the national government, the federal states and the communes continued until 1920, when the issue was decided in favour of centralization by the passage of the *Reichslichtspielgesetz*.

Films in the city

The question of whether there were local differences in film supply and/or audience preferences has also not yet been explored much. The Siegen research project provides unique tools for such an enquiry, because it not only documents the films on offer in Germany, but also a large sample of films actually screened. In Germany, the first permanent cinemas, the *Ladenkinos*, apparently showed the same types of films as itinerant cinemas did, mainly fictional ones, and both showed programmes comprising several short films, modelled on the programmes of urban variety theatres. Conversely, variety theatres, especially those attended by the higher classes, integrated mainly non-fiction films into their (predominantly live) shows.[34] The programme structure of short numbers of diverse content seems to have worked equally well for city and country, while the balance of entertainment and information in the programme depended on the exhibition context and the audience.

In the period after the turn to longer narrative films around 1910, questions regarding local differences are more difficult to answer. So-called *Monopolfilme* ('exclusives') were mainly intended for the *Kinopaläste*, large cinemas for sophisticated audiences in big cities, and were thus distributed in a limited number of copies at an above-average price. Small-town audiences did not get to see them immediately, if at all. Contemporary discourses seem to have exaggerated the cultural convergence between city and country brought about by cinema:

> We should not be indifferent about the kind of mental fare that is accepted and digested in German and neighbouring metropolises by the social trendsetters, or perhaps just the indiscriminate masses. The connections between these metropolises and our centres are so close, and the traffic is so smooth, that the 'hit' that only yesterday tickled the exhausted imagination of the urbanite already today is greeted with a knowing smile by the provincial, and tomorrow is gawked at as the latest fashion by the farmer's son.[35]

In retrospect, German discourses were often prescient in their predictions of the direction that the diffusion of cinema would eventually take, but because of their highly critical and conservative stance, they tended to overestimate its speed and consequences, and particularly its negative effects in displacing other forms of art and entertainment and its 'corruption' of youth. In general, the Siegen database suggests that there were large variations in the number and variety of films on offer, but that this did not depend solely on the size of the location or the degree of urbanization; whether variations were due to differences in the distribution-infrastructure, censorship practices or the degree of homogeneity of audience preferences requires more research (see Table 8.2).

Cities in film

Because of their relative accessibility and popularity, films offer an excellent source for studying the image of urbanization among the wider public. Analysing individual films can be of great value, if a systematic sample rather than a canonical one is chosen.[36] In fact, contemporaries already mused that scholars might find this interesting:

> Let us imagine the possibility that the cinematograph today were 100 years old rather than just more than 10, so that moving images of the good old days had been handed down to us. What a vivid image of the past these faithful documents would afford us! A single reel of film would reveal more than two dozen books based on laborious research and boring studies. Let us, for instance, take

TABLE 8.2 Number of different films screened between 1905 and 1914 in nine German locations

Big cities	Hamburg 447	Leipzig 416	München 350	Total 1213
Medium-size cities	Hagen 267	Görlitz 409	Würzburg 416	Total 1092
Small towns	Rendsburg 221	Pirna 74	Weiden 133	Total 428
				Total 2733

a street view. What insights would it offer into the past life and activities of a small town! It would not only show clothes and shoes, but also the manner of walking, the form of greeting, the social mores, the movements in the street – in short, everything that today we may conceive of only on the basis of extant paintings, engravings and descriptions.[37]

The city as topic does not seem to have been of as much interest as one might be led to expect by the modernity thesis: of the 45,000 films on offer in Germany between 1895 and 1920, only 324 contained the word *Stadt* ('city') in their title, and only 22 in the programme sample.[38] Film titles were especially important for early cinema, because little else of their content was communicated in advance, so they often referred to the central motif. Of course, films that did not deal with the topic explicitly would also have shown cities, for example as a backdrop to the plot. But it is reasonable to assume that films explicitly dealing with the topic presented the prevalent urban stereotypes, so at least for the purpose of constructing a meaningful typology this sample is useful as well as manageable.

The distribution of individual cities in the titles of supplied films is noteworthy: Paris appears in 312 titles, Berlin in 264, London in 83, New York in 79, Hamburg in 77, Munich in 51, Leipzig in 41, and Chicago in 21. Only a few instances appear in the programme-sample: from one of Chicago to 21 of the German capital Berlin (sometimes called *Chicago an der Spree*) and 42 of Paris. The production country was usually the home country of the respective city portrayed. With the exception of films of Berlin, views of particular German cities were almost always only shown in that city, indicating that film companies made city views to use as local films, which were immensely popular with audiences.[39] A contemporary comment on actualities is significant in this context:

> That today's periodically appearing actualities (*Gaumont-Woche, Pathé-Journal*) do not meet with greater success is due to their ... editorial rigidity. What in Paris may be ... a long leading article may be only of value as a brief notice in Berlin. An image of only local actuality may be of greater interest in Zurich or Krähwinkel [a fictional, generic name for a small town] than a world event dimmed by distance.[40]

Even fairground cinema, which was oriented towards transcultural entertainment and thus neither very interested in actualities nor able to supply them (because films were usually bought rather than rented), occasionally presented such 'self-images'. For example, a programme by the travelling showman Heinrich Hirdt at the Oktoberfest of 1906 was not only entirely made up of films of Munich, but even included a film of the Oktoberfest parade. Despite the transnational exchange of city images and early cinema's undoubted fascination with presenting a 'view of the world', these city films and their use reflect German audiences' strong interest in their own 'place' as well as their position on urbanization and cosmopolitanism.

The city films can be grouped under different types, which reflect the complex and contradictory attitudes towards urbanization.[41] Most of the non-fiction city films showed picturesque old cities (e.g. *Die alte italienische Stadt Pisa* [*The Old Italian City of Pisa*], 1913) rather than modern ones. These travel views usually portrayed cities as interesting, inviting places.[42] Even in the case of Berlin, the epitome of modernization, the old part of the city was shown as well as the new (e.g. *Eine Fahrt durch Berlin* [*A Journey through Berlin*], 1910). Berlin was usually less a topic for its own sake than for its technological innovations (e.g. *Eine Hochbahnfahrt durch Berlin* [*An Elevated Railway Joruney through Berlin*], 1910) or for events connected with the monarchy (e.g. *Originalaufnahmen vom Zarenbesuch in Berlin* [*Original Images of the Tsar's Visit to Berlin*], 1913) – reflecting the fact that Berlin was only a capital in some functions and that it was perceived differently by different social groups.[43]

Fictional city films dramatized the old moral contrast between city and country in a comic or tragic manner (e.g. *Möricke aus Neu-Ruppin kommt nach Berlin* [*Möricke from Neu-Ruppin Comes to Berlin*], 1911 and *Im Labyrinth der Großstadt* [*In the Labyrinth of the Great City*], 1912). Historical spectacle films often dealt with the decline and fall of (pre-modern) cities (e.g. *Trojas Fall/La Caduta di Troia* [*The Fall of Troy*], 1911, *Die letzten Tage von Pompeji/Gli Ultimi Giorni di Pompei* [*The Last Days of Pompei*], 1913, *Das Festmahl des Belsazar/Le Festin de Baltazar* [*The Feast of Belshazzar*], 1910). The tragic type set in the present was an explicit and very popular subgenre, the *Großstadtdrama* ('big city drama').[44] Since film dramas (especially 'exclusives') were predominantly shown in larger cities, it is significant that their audiences apparently took particular pleasure in horrific dramatizations of the actual or assumed vices of the cultural environment in which they were living.[45]

An allegorical and quite morbid connection between fiction and non-fiction, past and present was explicitly proposed in a description of a view of Messina in connection with the earthquake in 1909 (*Messina vor der Zerstörung* [*Messina before the Destruction*], 1909): 'Wonderful, original shots that visualize the glory and tragic destiny of magnificent cities'. The pronounced sensationalism of early cinema commonly underlay the different types of city films. In the sample, most films connected with Paris were either views of the city's 1910 flood or erotic films (e.g. *Pariserin im Bade* [*Parisienne Bathing*], 1897, still being shown ten years later). Adult programmes were sometimes even called *Pariser Abende* ('Parisian evenings'). As regular features, they seem to have been possible only in the anonymity of big cities; when travelling showmen presented them in small towns, they usually did so as their last show, so they could leave immediately afterwards in order to escape censure.[46]

The fact that the intended and perceived meanings of early films were often shaped by the context of the programme and the social space in which they were viewed can be poignantly illustrated with the following example. On 16 October 1901, the Hansa-Theater, a high-class variety theatre attended mostly by the moneyed bourgeoisie, showed the film programme illustrated in Figure 8.3.

The film of the demolition of the Star Theatre has often been interpreted as an anticipation of the avant-garde 'city symphonies' of the 1920s, because of its anti-realistic effects, such as using time-lapse photography and showing the demolition in reverse. But note that the water sports film was shown with the same gimmick.[47] In the context of the programme's presentation of the funeral procession, turning back

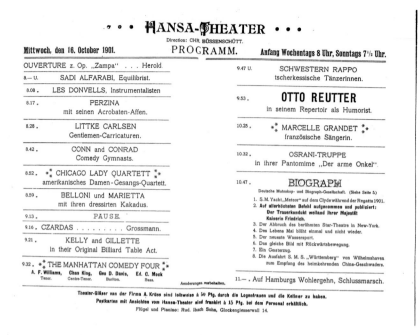

FIGURE 8.3 1901 film programme of a high-class variety theatre in Hamburg

time seems to imply nostalgia rather than modernization. In Germany, early films were situated at least as much in the 'old world' as in the 'new', and the new media technology did not have to be in accord with the experience of modernity, but could just as well express the views of and yearning for 'the good old days', even in a contradictory and reactionary manner.

Conclusion

The relationship between early cinema, urbanization and modernization is much more complex and indirect than contemporary critics and the current 'modernity thesis' seem to imply. The claim of an essential affinity between 'urban perception' and cinema, of a 'homology between cinematic spectatorship and urban experience', is particularly questionable.[48] Regardless of their attitude towards modern city life, not all urbanites attended early cinema (many workers had neither the time nor the money to do so), nor was 'cultural conditioning' through the modern urban environment necessary to understand and enjoy cinema, as the new technology's success in rural areas demonstrates. This notion was, indeed, already being questioned on an empirical basis by at least one German contemporary: in 1912, Emilie Altenloh showed in her sociological study of audiences in Mannheim and Heidelberg that there were no differences in the frequencies of cinemagoing between natives and long-time inhabitants of the city, nor between recent migrants and commuters from the country.[49]

Whether individuals attended cinema depended on two major factors: opportunity and preference. Of course, apart from socio-economic factors such as income and leisure time, which were influenced by modernization, urban audiences had more opportunities than rural audiences. But media preferences were influenced by much more complex socio-psychological factors, both transcultural as well as culturally specific, than the concept of an 'urban perception' allows for. In the short as well as the long term, cinema became a highly successful modern medium, because despite its technological and economic standardization, its institutions were able and willing to adapt to a wide variety of contexts and tastes in order to be enjoyable for many different people, rather than for only one particular 'perception'.

Notes

1 My thanks to Joseph Garncarz, Martin Loiperdinger and Klaus Kreimeier for comments, and Michael Ross for help with the database queries.
2 H. Kienzl, 'Theater und Kinematograph', *Der Strom* 1, 1911, 219–21. Translations are mine unless otherwise stated.
3 See e.g. J. Donald, *Imagining the Modern City*, London: Athlone Press, 1999, and G. Vogt, *Die Stadt im Film: Deutsche Spielfilme 1900–2000*, Marburg: Schüren Verlag, 2001. For a general overview of cultural theory on the modern city, see e.g. D. Frisby, *Cityscapes of Modernity: Critical Explorations*, Cambridge: Polity Press, 2001.
4 Notable exceptions for the German context are e.g. Brigitte Flickinger, 'Zwischen Intimität und Öffentlichkeit: Kino im Großstadtraum: London, Berlin und St. Petersburg bis 1930', in C. Zimmermann (ed.) *Zentralität und Raumgefüge der Großstädte im 20. Jahrhundert*, Stuttgart: Steiner, 2006, pp. 135–52; C. Zimmermann, 'Städtische Medien auf dem Land: Zeitung und Kino von 1900 bis zu den 1930er Jahren', in C. Zimmermann (ed.) *Die Stadt als Moloch? Das Land als Kraftquell? Wahrnehmungen und Wirkungen der Großstädte um 1900,* Basel/Boston/Berlin: Birkhäuser, 1999, pp. 141–64. Particularly thorough examples of local studies are D.H. Warstat, *Frühes Kino der Kleinstadt*, Berlin: Spiess, 1982, and A. Paech, *Kino zwischen Stadt und Land: Geschichte des Kinos in der Provinz: Osnabrück*, Marburg: Jonas, 1985. See also J. Steffen, J. Thiele and B. Poch (eds.) *Spurensuche: Film und Kino in der Region*, Oldenburg: BIS, 1993.
5 For work on the Siegen research project 'Industrialization of Perception' (2002–9), on which the following is based, see e.g. M. Loiperdinger (ed.) *Travelling Cinema In Europe: Sources and Perspectives*, Frankfurt a.M./Basel: Stroemfeld/Roter Stern, 2008; J. Garncarz, *Maßlose Unterhaltung: Die Etablierung des Kinos in Deutschland, 1896–1914*, Frankfurt a.M./Basel: Stroemfeld, 2010; M. Ross, M. Grauer and B. Freisleben (eds) *Digital Tools In Media Studies: Analysis and Research: An Overview*, Bielefeld: transcript, 2009; A. Ligensa and K. Kreimeier (eds) *Film 1900: Technology, Perception, Culture*, New Barnet: John Libbey, 2009.
6 See e.g. A. McElligott (ed.) *The German Urban Experience, 1900–1945: Modernity and Crisis*, London/New York: Routledge, 2001; and Th. Elsaesser and M. Wedel, 'Mimesis, Mimicry and the Metropolis: a Century of Berlin's Multi-National Film Cultures', in G. Weiss-Sussex and U. Zitzlsperger (eds) *Berlin: Kultur und Metropole in den zwanziger und seit den neunziger Jahren*, München: Ludicium, 2007, pp. 136–52.
7 On the 'modernity thesis' debate in early cinema studies, see e.g. B. Singer, *Melodrama and Modernity: Early Sensational Cinema and Its Contexts*, New York: Columbia University Press, 2001; and F. Kessler, 'Viewing Change, Changing Views: The "History of Vision"-Debate', in A. Ligensa and K. Kreimeier (eds) *Film 1900: Technology, Perception, Culture*, pp. 23–37.
8 A. Lees, *Cities Perceived: Urban Society in European and American Thought, 1820–1940*, New York: Columbia University Press, 1985.

9 'Kinematographenbesitzer und Lehrerschaft', *Allgemeine deutsche Lehrerzeitung* 66, 1914, 16–18.

10 A. Hellwig, 'Die Beziehungen zwischen Schundliteratur, Schundfilms und Verbrechen: Das Ergebnis einer Umfrage', *Archiv für Kriminalanthropologie und Kriminalistik* 51, 1913, 1–32.

11 See e.g. A. Haller, 'Seen Through the Eyes of Simmel: The Cinema Programme As a "Modern" Experience', in *Film 1900: Technology, Perception, Culture*, pp. 113–24.

12 For a more thorough discussion, see e.g. B. Singer, *Melodrama and Modernity*, Chapter 1: 'Meanings of Modernity', pp. 17–36.

13 Expressed by the title of: Th. Elsaesser and M. Wedel (eds) *Kino der Kaiserzeit: Zwischen Tradition und Moderne*, Munich: edition text + kritik, 2002. For a contemporary articulation, see e.g. the polemic analysis of Berlin's architecture by W.Rathenau, 'Die schönste Stadt der Welt', in *Impressionen*, Leipzig: Hirzel, 1902, pp. 140–62.

14 B. Singer, 'The Ambimodernity of Early Cinema: Problems and Paradoxes In the Film-and-Modernity Discourse', in *Film 1900*, pp. 37–53.

15 The databases are available online at http://www.fk615.uni-siegen.de/de/teilprojekt.php?projekt=A5.

16 For a detailed description, see J. Garncarz and M. Ross, 'Die Siegener Datenbanken zum frühen Kino in Deutschland', *KINtop: Jahrbuch zur Erforschung des frühen Films* 14/15, 2006, 151–63.

17 S., 'Der Kinematograph', *Komet* 600, 19 September 1896, 3.

18 Joseph Garncarz assumes that travelling played a lesser role in the USA compared to town hall players (who used fixed, often multi-purpose sites rather than transporting cinema buildings). See J. Garncarz, *Maßlose Unterhaltung: Die Etablierung des Kinos in Deutschland, 1896–1914*, Frankfurt a.M./Basel: Stroemfeld, 2010; and compare this with Ch. Musser, *High-Class Moving Pictures: Lyman H. Howe and the Forgotten Era of Traveling Exhibition, 1880 – 1920*, Princeton, NJ: Princeton University Press, 1991; and Kathryn H. Fuller, *At the Picture Show: Small-Town Audiences and the Creation of Movie Fan Culture*, Washington: Smithsonian Institution Press, 1996.

19 See J. Garncarz, 'Perceptual Environments For Films: The Development of Cinema In Germany, 1895–1914', *Film 1900*, pp. 141–51.

20 M. Hölz, *Vom 'Weißen Kreuz' zur roten Fahne: Jugend-, Kampf- und Zuchthauserlebnisse*, Berlin: Malik-Verlag, 1929, p. 22.

21 On fairs in Germany, see e.g. L. Abrams, *Workers' Culture in Imperial Germany: Leisure and Recreation in the Rhineland and Westphalia*, London/New York: Routledge, 1992.

22 Garncarz, 'Perceptual Environments For Films'. For the comparison of cinemas and department stores, derived from Walter Benjamin's 'arcades project', see e.g. A. Friedberg, *Window Shopping: Cinema and the Postmodern*, Berkeley, CA: University of California Press, 1993.

23 T. Coles, 'Department Stores as Retail Innovation in Germany: A Historical-Geographical Perspective on the Period 1870–1914', in G. Crossick and S. Jaumain (eds) *Cathedrals of Consumption: The European Department Store, 1850–1939*, Aldershot: Ashgate, 1999, pp. 72–97.

24 C. Müller and H. Segeberg (eds) *Kinoöffentlichkeit (1895–1920)/Cinema's Public Sphere (1895–1920)*, Marburg: Schüren, 2008.

25 Flickinger, 'Zwischen Intimität und Öffentlichkeit', p. 145.

26 See e.g. L. Abrams, *Workers' Culture in Imperial Germany*; and G. Kilchenstein, *Frühe Filmzensur in Deutschland: Eine vergleichende Studie zur Prüfungspraxis in Berlin und München 1906–1914*, Munich: Diskurs-Film-Verlag Schaudig und Ledig, 1997.

27 Garncarz, 'Perceptual Environments For Films'. Efforts by German cinema reformers to re-establish itinerant cinema for educational purposes were unsuccessful. Interestingly, the idea of a community cinema already arose in this context. See e.g. W. Warstat, *Kino und Gemeinde*, Mönchengladbach: Lichtbilderei, 1913.

28 R. May, 'Filmpolitik', *Vossische Zeitung* (morning edition), 515 (9 October 1912).

29 C. Ross, 'Mass Culture and Divided Audiences: Cinema and Social Change in Inter-War Germany', *Past & Present* 193, 2006, 157–96.

30 H. Gathmann, 'Erster kinematographischer Instruktionskursus im Gemeinde-Lichtspielhaus in Eickel i.W.', *Die Hochwacht* 3, 1913, 163–65.
31 According to industry data quoted in e.g. P. Samuleit, 'Die Kino-Frage vor der deutschen Lehrerschaft', *Volksbildung* 42, 1912, 226–28. Germany had a population of about 65 million at the time.
32 Steffen, Thiele and Poch (eds) *Spurensuche: Film und Kino in der Region*.
33 Kilchenstein, *Frühe Filmzensur in Deutschland*.
34 See Garncarz, 'Perceptual Environments For Films'.
35 B.H.A. Schmitz, 'Kino und Großstadtjugend', *Die Hochwacht* 4, 1913; 26–30.
36 For more detailed film analyses of early city representations see e.g. U. Jung, 'Local Views: A Blind Spot In the Historiography of Early German Cinema', *Historical Journal of Film, Radio and Television* 22, 2002, 253–73; N. Verhoeff and E. Warth, 'Rhetoric of Space: Cityscape/Landscape', *Historical Journal of Film, Radio and Television* 22, 2002, 245–51; L. Nead, 'Animating the Everyday: London On Camera c. 1900', *Journal of British Studies* 43, 2004, 65–90, 161–62; F. Kessler, 'Comment cadrer la rue?', in E. Biasin, G. Bursi and L. Quaresima (eds) *Lo stile cinematografico/Film Style,* Udine: Forum, 2007, pp. 95–101.
37 Ludwig Brauner in *Der Stadtverordnete*, quoted in 'Kinematographische Archive', *Österreichischer Komet* 10, 1912, 4. Since the 'visual turn', scholars outside film studies are increasingly discovering films as historical documents; see e.g. the 'visual sociology' of M. Horwitz, B. Joerges and J. Potthast (eds) *Stadt und Film: Versuche zu einer visuellen Soziologie*, Berlin: WZB, 1996.
38 See e.g. G. Bruno, *Streetwalking On a Ruined Map: Cultural Theory and the City Films of Elvira Notari*, Princeton, NJ: Princeton Unversity Press, 1993.
39 M. Loiperdinger, '"The Audience Feels Rather At Home": Peter Marzen's "Localiziation" of Film Exhibition in Trier', in F. Kessler and N. Verhoeff (eds) *Networks of Entertainment: Early Film Distribution 1895–1915,* Eastleigh: John Libbey, 2007, pp. 123–30.
40 A.J. Storfer, 'Die Zukunft der kinematographischen Zeitung', *Kinematograph* 261, 27 December 1911.
41 We were not able to identify all the films in our programme database. That is why some of the films mentioned here may be foreign productions despite the German title.
42 Interestingly, German cities had overtaken spa-towns in contemporary travel statistics. See McElligott, *The German Urban Experience, 1900–1945*, p. 150.
43 See D. Briesen, 'Berlin – Die überschätzte Metropole: Über das System deutscher Hauptstädte zwischen 1850 und 1940', in G. Brunn and J. Reulecke (eds) *Metropolis Berlin: Berlin als deutsche Hauptstadt im Vergleich europäischer Hauptstädte 1871–1939*, Bonn/Berlin: Bouvier, 1992, pp. 39–78. Briesen uses quantitative indicators to argue that Berlin was only first among equals in many respects, and that this did not change by becoming a formal and modernized capital. The positive perception of Berlin's techno-logical innovations among the moneyed bourgeoisie is also reflected in the popular press, e.g. the magazine *Gartenlaube*. See K. Dördelmann, 'Die Darstellung Berlins in der populären Zeitschriftenpresse, 1870–1933', in G. Brunn and J. Reulecke (eds) *Metropolis Berlin*, pp. 127–50.
44 See E. Altenloh, *Zur Soziologie des Kino*, Leipzig: Spamer, 1913, p. 85.
45 Such depictions were also popular in written form, e.g. the *Großstadtdokumente*-Series, edited by Hans Ostwald, a project between literature, journalism and sociology. See e.g. P. Fritzsche, 'Vagabond in the Fugitive City: Hans Ostwald, Imperial Berlin and the *Großstadt*-Dokumente', *Journal of Contemporary History* 29, 1994, 385–402. Several of the authors involved even wrote film scripts.
46 See Ch. Junklewitz, *Erotik im frühen Kino: Ästhetik und kulturelle Praxis*, unpublished M.A. thesis, University of Cologne, 2004.
47 J.-Ch. Horak, 'Auto, Eisenbahn und Stadt – frühes Kino und Avantgarde', *KINtop: Jahrbuch zur Erforschung des frühen Films* 12, 2003, 95–119.
48 Donald, *Imagining the Modern City*, p. 64.
49 See E. Altenloh, *Zur Soziologie des Kino*, p. 77.

9

DIAGNOSIS: 'FLIMMERITIS'

Female cinemagoing in Imperial Germany, 1911–18

Andrea Haller

Introduction

In his 1918 book *Flimmeritis: What everybody needs to know about cinema*, film journalist Egon Jacobsohn defined *Flimmeritis* as an 'undeviating desire to become a movie star' and as 'a modern epidemic plague, that all of a sudden afflicts normally pretty harmless and sensible citizens and causes boisterous agitation and dangerous insanity'.[1] As early as 1917, Jacobsohn, who was the editor of the German fan magazine *Illustrierte Kinowoche,* had already published an article in which he made fun of the numerous German women who were seized by *Flimmeritis.*

This little article provoked an energetic discussion among the female readers of the magazine. In numerous letters to the editor they debated the legitimacy of German women going to the cinema and adoring movie stars. While some women held that it was perfectly acceptable to dote on movie stars and aspire to become a movie star oneself, others responded that the degree of adoration and the attempts of thousands of 'normal' girls to become movie stars sometimes became ridiculous. This discussion indicates the controversial nature of female moviegoing at the end of the German empire, even within the community of female cinemagoers. But what made it so controversial?

In her 1990 article on cinema and the public sphere, Miriam Hansen has pointed out that female movie patronage, which included the fan activities that accompanied attending the movies, has to be seen in the wider context of a discourse about the transformation of the public sphere in the age of modern mass media. Hansen argues that during the early twentieth century the formerly masculine and bourgeois public sphere was in the process of becoming increasingly feminized.[2] With its egalitarian appeal, cinema became the perfect place for women to step out from their domestic environment and enter the new public sphere. Cinema threatened to blur the boundaries between public and private, and made women literally visible, since they had to step out of the house and walk along the streets to enter the cinema. The discourse on female cinemagoing can, therefore, also be read as a discourse on visibility: on how women became visible in public and how they were visible in the auditorium. It can be read as a discourse on what was not visible in the dark auditorium and on what was suspected to happen there, as well as on what women could have watched on screen and on what they should not have seen there. Focusing on this concept of 'visibility', I would like to explore the discourse on female moviegoing in Imperial Germany, insofar as it can be reconstructed by studying the writings of the cinema reform movement and contemporary trade papers, as well as fan magazines and women's and fashion magazines that deal with female movie patronage from different perspectives.[3]

A sociology of the audience

In any discussion of movie patronage in Imperial Germany, Emilie Altenloh's 1913 study of contemporary cinema audiences is a central point of reference.[4] Altenloh's *A sociology of the cinema audiences* was a groundbreaking study because it provided some of the first statistical data about cinemagoing in the German *Kaiserreich*. One might question the methodology, but as Frank Kessler and Eva Warth have argued, her study 'clearly indicates that early cinema audiences were a much more complex phenomenon than contemporary reformists or later film historiographical discourses were prepared to see them as being'.[5] Unlike her contemporaries who often disapproved of or simply neglected the presence of women in the cinema, Altenloh not only recognized that women from all social classes constituted an important part of the audience, but particularly stressed women's enthusiasm for cinema. For working-class women, going to the cinema quickly became a habit and 'an important component' in their existence, she wrote. In the cinema, they 'live in another world, a world of luxury and excess which makes them forget their dull daily routine'.[6] Shop girls, secretaries and other workers subsumed by Altenloh into the category of female clerical assistants usually went to the better and more elegant cinema houses, and they liked to watch love-stories, films depicting the glamour of cosmopolitan metropolitan circles and especially films with 'content that they can most easily relate to their own lives and circumstances'.[7] According to Altenloh's findings, women from the upper classes went to the cinema even more frequently than workers and female clerical assistants, because their time was not limited by any occupation. For them, cinema offered a good means to bring some 'sensational stimuli' and some adventurous diversion into their carefree but somewhat boring everyday lives, she argued. They liked to watch films involving the conflicts of a woman's life, but Altenloh also suggested that they otherwise shared the tastes of working women and shop girls. A more general reason why cinema was so popular was the fact that 'both the cinema and those who visit it are typical products of our times, characterized by constant preoccupation and a state of nervous restlessness'.[8] Cinema, as a specifically modern medium involving 'devotion to and immersion in the present' allowed women to feel part of the modern world.[9]

The cinema reform movement and female moviegoers

Like Altenloh, the predominantly male writers of the cinema reform movement recognized that women were to be found in the new permanent movie houses that opened in large numbers around 1907. Unlike Altenloh, they regarded cinema's modernity and its appeal to all classes and both sexes with unease. The reformers initially directed their criticism against the exhibition sites, and soon began to attack women's presence in these venues. Their deeper concerns about the presence and visibility of women in the auditorium were barely hidden behind their doubts about women's safety there. Their concerns were rooted in a larger discourse about women's increasing presence in public and on the streets, as they became an important part of the developing consumer society.[10] In his 1913 pamphlet against the cinema, *Der Kino*, social reformer Victor Noack wrote that cinema 'has temptingly facilitated sexual attacks on children and women'.[11] His report of these attacks implied that sexual assaults on women in the cinema were a daily occurrence. When writing about cinema as a dangerous place, reformers mainly had young unchaperoned girls in mind. Austrian politicians for example claimed that not only teenagers 'of both sexes sit there crowded together promiscuously' but that older men in particular took the opportunity to approach young girls.[12] The trade paper *Der Kinematograph* that reprinted these accusations answered in a sardonic tone:

> The older and younger gentlemen who approach the female sex, especially the young teenagers (*Backfische*), can be found not only in the cinema, but throughout Vienna. Besides, – as far as we know – these 'sexual gourmets' mostly originate from the 'better' circles that try to raise the morality of the people (*Volkssittlichkeit*). Therefore cinema is not more dangerous than the legitimate theatres, concert and dance halls in general.[13]

Such accusations persisted in the next decade: in 1920, a speaker in parliament claimed that half the underage women in childbed in a clinic in Munich had met their seducer in the cinema.[14] The internal logic behind these arguments was paradoxical: it was understood to be dangerous for girls and women to become visible, because everybody could see them when the lights were on, but it was apparently considered even more dangerous that nobody could see them when the lights were turned off. Although there may have been little by way of evidence to support these suggestions, they became a kind of common knowledge about cinema, used to discredit it as a legitimate form of leisure activity for young girls and women. As a new social space, the cinema enabled women to move from private space to a public environment in which they became equal members of a wider audience. As Sabine Hake notes, the new freedom that cinema offered to contemporary women had to be stigmatized as dangerous, because women in the cinema challenged traditional notions of femininity.[15]

Cinema was accused of causing moral and even mental harm as well as physical harm. First, the structure and composition of the programme, its variety of attractions and the emotional and visceral stimuli put together in one show made it an extremely modern medium. It was held responsible for the hysterical seizures and hallucinations that women were allegedly afflicted with, and for modern ailments such as neurasthenia, a nervous disease resulting from an 'overload' of the brain.[16] As Scott Curtis has argued, concerns about spectators' physical health transformed themselves into questions of moral health as soon as reformers turned against the 'trashy film' (*Schundfilm*).[17] Most of the 'cine-drama' (*Kinodramen*) – both social dramas and sensational or *Sitten*-dramas that were vilified as 'trashy films' – featured female leading characters and narrated stories from the realm of female experience. Heide Schlüpmann has even suggested in 1911 that there was a 'secret conspiracy' between German cinema and its female audiences.[18] Reformers accused the *Schundfilme* of giving the female audience wrong ideas about both social and sexual issues.[19] The reform discourses reveal a fear of scopophilia on the part of women, and a more general fear of the visualization of women themselves, in the auditorium and on the screen. As Hake puts it, 'The image of women seeing rather than being seen contributed to the impression that cinema had a liberating effect on women.'[20]

The numerous reports of sexual assaults in the cinema and of the threats it posed to women's physical, mental and moral health do not, however, tell us very much about the actual risks in cinemagoing. They merely reflect male apprehensions about the appearance of women in the public space of the cinema and their changing role in modern society. The fact that women themselves did not feel physically or morally endangered in the cinema but instead embraced the new opportunity to go out and partake in modern life may have increased their apprehensions.

Women in the cinema

Sources from the trade press, fan and women's magazines show that, unlike the reformers who worried about women's safety in the cinemas, women perceived cinema as a 'safe' place in which they could take a rest from the rush of the streets, from their exhausting shopping tours or their everyday work. They perceived cinema as a place where women could enter without *Schwellenangst* (fear of entering a place or of something new), where they could go alone and without spending a lot of money. The 1912 article 'Cinema characters' sees the 'Mutti' or the married woman from the countryside, as a typical cinema type:

> Day in, day out cooking, washing, ironing, patching and darning socks. At last she can escape and drives to the city, runs from Alsberg to Tietz [two big shopping centres], and to the brothers Müllern. The packages become bigger and bigger. She is hungry. She can afford a coffee for 20 Pfennig. But where should she drink it? On the street? Her arms are full, and then the rude boys … In a café? It is too elegant there. The only way open is the cinema. It is dark there. And there is something to see. There she can sit safely; she puts her packages besides her, crackles with her cake cornet and chews with both cheeks.[21]

And she is so moved by the pictures that she nearly drops her pastry. This little *vignette* not only makes it clear that going to the cinema was embedded in everyday practices like shopping and running errands, but also suggests that women enjoyed being affected and venting their emotions in the cinema. In 1915, journalist Anna Knust described in her 'Cinema letter from a provincial town' how a female cinema regular spontaneously cried out 'Thank God', when the young and beautiful heroine was saved from the burning house. Female moviegoers did not sit motionless and speechless, but became active and verbalized their emotions impulsively.[22] As Altenloh suggested, being moved and verbalizing these emotions was not a matter of social class or rural or urban affiliation, and women weeping in the cinema had become a fixed topos by then. Resi Lange, an actress and elocutionist who frequently wrote about cinema, characterized the attendees of an elegant cinema in Berlin West in 1913:

> The junior clerk and the shop girl. Underneath her white blouse a sensing heart is beating at the strange fates of the cinema heroes, and the junior clerk lays his cold hand on her hand with the golden ring, and he says in a shaky-sentimental tone: 'Do you like it, miss?' And while she nods 'Yes', a tear falls from her lashes onto his cold hand, a silent tribute of emotion.[23]

The *Kinematograph* cited an old miner's wife with the words: 'That was so moving that my handkerchief was totally wet when I came home. … When I want to have a really good cry I go to the cinema.'[24] Weeping in public could be seen as an extreme version of turning the private into public, of showing publicly what had previously been a solely private matter. And, as a survey on the cinema in post-war Britain showed, weeping in the cinema was not a private but a collective experience, especially for women.[25]

It was not only emotion that drew women into the cinema. The diversity of the short film programme attracted female audiences, who embraced the various attractions on offer without fainting or becoming hysterical, as the reformers had feared. One essayist reported that he attended the lavish opening of the new *Nollendorf-Theater* in Berlin for the premiere of the 'giant film' *Quo Vadis*. While he was impressed by the film's artistic quality, the big orchestra and the theatre-like architecture of the cinema building, his female companion complained, 'Instead of this big antique thing I would have preferred another program: the latest fashion, the *Kaiser*, Lehmann, Asta Nielsen, the handsome Max [Linder] or "Sinful Love." Something like that.'[26] The bourgeois male author did not forget to mention that her taste represented the taste of the masses, expressing the then commonplace view that 'the masses' were implicitly feminine.[27]

Male and female ideas about an enjoyable cinema experience seemed to conflict. In 1913, the fan paper *Illustrierte Kinowoche* published an essay by Poldi Schmidt about going to a cinema with 'Irma, the cinema girl' in which he explained why he would never do that again.[28] While he wanted to watch the movies his girl-friend was constantly chatting. She talked about the actors, the film companies and even revealed the ending of the film. He was disenchanted, he complained, because she constantly disturbed his concentration. Reading between the lines, the account reveals that the girl read the films in her own way, laughing when others were crying, for example. Her way of acting towards the films displayed her media competence and viewing skills, which helped her understand how the film stories were constructed. She was a self-confident film connoisseur who knew the conditions of production and reception, an active consumer and by no means the overwhelmed and unstable object portrayed by reformers. While Schmidt took up a viewing position that would become typical for classical Hollywood cinema, she retained the mode of reception typical for early cinema.

There were other reasons why women liked to go to the cinema. In his sketch in the *Lichtbild-Bühne* 'My wife in the film theatre', Robert Wilke sardonically described a visit with his wife to the cinema.[29] According to his wife, one reason for going to the pictures rather than the theatre was that she did not need to dress up. Then he describes Mrs Wilke – who only needs 'two minutes' to get ready – changing her clothes several times and posing in front of the mirror to find the right outfit for the cinema. His wife actually loved to go to the movies in order to show off her latest dress or latest hairdo in public. The cinema was also the perfect place for her to identify the latest trends and define her status in the world of fashion. She delivered long monologues when the latest hat models were presented on screen, and when she was bored by the film, she commented on the hats of the other women in the audience and criticized the hairdo of the 'lady in the third row'. This behaviour clashed with Wilke's view of the purpose of going to the cinema, which was to watch films. As Shelley Stamp says about American female audiences, 'Women's tendency to parade themselves at leisure outings shifted visual attention away from the screen and onto the circulation of gazes in lobby areas and entranceways.'[30] Women were not always 'ideal' spectators who vanished in the dark auditorium. Instead, they made themselves visible and audible. 'Women were clearly not seamlessly integrated into the social space of the theatres; in fact, their increasing visibility was frequently more troublesome than beneficial during these years.'[31]

By 1915, cinema owners were being encouraged to educate their clients as ideal spectators, adopting attitudes of reception suitable to the upcoming longer feature film. In his article 'Educate the cinema audience', Albert Walter claimed that people should behave in ways that were 'suitable for theatrical performances', concentrating 'on the films projected on screen' and not commenting on the films, 'the actors, the weather, the children, about the latest quarrel with the cook or about the new blouse of Mrs Müller and the hat of Ms Lola'.[32] These topics make it clear that Walter believed that the female audience was in need of education, and Schlüpmann has argued that this 'education' was first and foremost a matter of eliminating utterances so that viewers could focus only on the films that now controlled the way in which they should be received. Female moviegoers and the audience in general should only witness (*miterleben*), not experience themselves (*erleben*) any more.[33] They should again become invisible and inaudible.

Walter also appealed to the female audience to dress properly: above all they should take off the oversized hats that blocked other people's view. At first this seems to be just an amusing piece of marginalia in film history, but looking more closely it reveals something about male attitudes to the female cinema audience and more generally to women in public. Complaints about the 'hat problem' probably form the most frequent commentary on female visibility in the cinema. The topos of a lady blocking the view of another, predominantly male, patron, had already become the subject of numerous essays, films and even poems by 1915. These fashionable large hats decorated with feathers, ribbons and even fruit were the clearest sign of women's presence in the cinema. They could not be overlooked by anybody; neither by women enviously examining the hat of the others, nor by men. Demands that women be forced to take off their hats might be understood as an attempt to limit female 'visibility' in the cinema and the public realm. Catherine Russell states that the debate about the 'big hats' externalized deeper 'anxieties brought about by the gendered transformation of the public sphere'.[34]

Women as movie fans

Along with the growing interest in cinemagoing, women actively participated in the process of reception outside the cinema as movie fans. From 1913 on, fashion and women's magazines like *Die Dame* or *Elegante Welt* frequently reported on films and their stars. The first movie fan magazine, *Illustrierte Kinowoche*, was also launched in 1913. The magazine's format, with its numerous reports on fashion, the life and work of stars, and especially the 'letters to the editor' section and the character of the advertisements, indicate that the majority of the readers were women. Female readers were addressed in special sections such as the column, 'I know everything. Letters from the editor's desk', presenting the latest news and gossip from the film business.[35] This weekly column took the form of a letter either to a fictitious mother, ('Dear Madam'), or her daughter ('Dear Mademoiselle'), suggesting that a love of the movies was not a matter of age.

Female readers not only consumed the latest news about film and cinema, but also became active themselves by sending in letters asking for the addresses of actors and actresses, demanding more information about their private lives and careers. They participated in polls about the most popular stars, sent in film ideas and film scripts and founded clubs for cinema lovers.[36] Reading fashion magazines like *Die Dame* and *Elegante Welt* indicates that they considered cinema to be part of a sophisticated and modern lifestyle, in which film stars increasingly served as lifestyle and fashion role models. These magazines often illustrated their articles on beauty, lifestyle and society matters with images of movie stars who served as the perfect embodiment of all that was fashionable, up-to-date and modern.[37] Responding to readers' demands, the magazines published film stills and discussed the costumes of the latest films.[38] The actresses themselves became fashion role models, and fashion magazines began to feature photo sessions with film actresses presenting the latest fashion and hat trends. In 1917, *Die Dame* reserved an entire page for Henny Porten presenting the latest hats.[39] Fan and fashion magazines offered their readers the opportunity to acquaint themselves with the life of a film star, reporting on Henny Porten's sporting habits or the holiday activities of different stars.[40] They illustrated stories about the domestic life of Asta Nielsen and Erna Morena with semi-private photos of the actresses in their home and with their children.[41] The demand for photos of the stars beyond 'official' images and film stills was fed with snapshots of the stars at social events such as horse races or theatre premieres.[42]

Changing over to the other side of the silver screen

The best way to be part of the business was to become an actress oneself. According to the numerous enquiries in trade papers and in fashion and fan magazines, a lot of German women and girls dreamed of being a film star. Female movie fandom culminated in the desire to become an actress, to step out from the dark auditorium into the studio lights and onto the silver screen, to acquire ultimate visibility and publicity and to become a truly modern woman.

From 1912 on, when the discussion about female cinema patronage was still in full sway, women bombarded editorial offices with questions about how they could become cinema actresses.[43] The trade press and fashion and fan magazines warned that it was not easy. The *Elegante Welt*, for example, answered to a request of L.K. from Grünstadt: 'As far as our experience goes we can only strongly advise you against taking up the profession of a film actress. Only a few very talented women made good as an actress.'[44]

By featuring numerous articles and stories on young girls trying to become film actresses, the *Illustrierte Filmwoche* aroused interest and raised hopes that readers might be discovered by film producers. At the same time, they warned their readers with stories in which people tried to take advantage of the girl's desire to succeed in the film business and in which the girls never fully succeeded in becoming a star. In particular, these stories suggested that the so-called 'cinema schools' (*Kinoschulen*) had become a major problem, attracting young girls with the promise of authorized qualifications as cinema actresses. In fact, these 'schools' profiteered by squeezing money out of the girls. The 1918 *Illustrierte Filmwoche* sketch 'Die Verkäuferin' tells the story of young Grete Müller who has always been a decent and reliable girl who works in a shop selling linen goods. Her only fault is that she suffers from *Flimmeritis*. She quits her job and pays 200 Marks a month to attend a *Kinoschule* with twenty other girls, shop and servant girls, daughters from the nobility, and 'talents' from the countryside. When they finish their studies, Grete tries to get work at a film company but is rejected everywhere. The story ends: 'Today Grete stands again behind the counter and sells linen goods. Just as it used to be, but without her bankbook and without *Flimmeritis* … '[45]

All these reports about impostors and the failure of so many girls did not prevent female film enthusiasts from dreaming of a career in the film business and continuing to ask for advice on how to become a movie star. In 1917, the *Elegante Welt* announced that 'We are no longer able and willing to answer the continual question "How can I become a cinema actress?"!'.[46] Damning the *Kinoschulen* and warning women and girls against seeking a career in the film business can be seen as another strategy to inhibit women's increasing presence and visibility in the public sphere.

Female movie attendance during the First World War

By 1914, the first wave of indignation against female movie attendance had calmed down, but, as Schlüpmann has observed, 'against the background of the up-coming war the dominant presence [of women] at the sites of scopophilia was again perceived as a provocation'.[47] Female visibility in the cinema continued to be viewed as a problem: patriotic German women should stay at home, fighting invisibly for Germany's welfare on the 'home front' (*Heimatfront*). In practice, however, women became an even more important part of the audience.[48]

The fact that women visited the cinema in such large numbers, looking for distraction from wartime conditions, provoked strong reactions. 'Do people who stay at home, especially women', the newspaper *Der Reichsbote* asked, 'really have the time and interest to go to the cinema?'[49] Women 'staying at home' definitely had both time and interest, despite the disapproval of their fathers, husbands and politicians. In response, the public authorities of numerous German cities deprived several 'soldiers' wives' (*Kriegerfrauen*) of their national financial support when they had been caught visiting the cinema, reproaching them for wasting their money on 'needless' leisure activities, accusing them of 'being idle and leading an amoral lifestyle', and denouncing their 'lack of domesticity and frugality'.[50] Cinema continued to be seen as a major threat to women's moral sense, patriotism, and self-sacrifice during the war. But German women firmly defended their right to go to the cinema. Hilda Blaschitz answered the *Bild & Film*'s question 'Now, and to the cinema'!? with a clear 'Yes', arguing that cinema had a particular *raison d'être* in times of war because it offered distraction and comfort to women who had to carry the burden of maintaining the everyday life at the *Heimatfront*.

> I see two office girls. They now have to work two-fold and three-fold … Small is their salary and great their work and their heads are exhausted; and they worry: about their sweethearts, about their brothers. For some short moments cinema is a comforter for them.[51]

Conclusion

The discourse on female movie patronage in Imperial Germany must be seen in the wider context of changes in the public sphere in the challenges that modernity brought into being. This is further complicated by the difficulty in differentiating between the historical figure of the female moviegoer and her discursive construction. Cinema not only allowed women to step out into the public sphere. It also publicized their conditions of life, showing on screen what had previously been private and domestic and offering the female moviegoers a hitherto unavailable chance to 'see themselves', their everyday life, their milieu and their experiences. In a book on cinema in Imperial Germany published in 1912, Nanny Lutze observed that 'in nearly all the pictures a woman is playing the leading role! … It is a piece of life; an extract, an episode – maybe!? – from our being!'[52] She continued: 'Warm and red the blood rushes to our cheeks – we get cold and warm – we frighten and shudder – we tremble and despair – we cheer and weep with these … schemes on the silver screen.' As well as offering German women only some hours of relaxation and entertainment, cinema also gave them the chance to become visible and to see what could not be seen before.

Notes

1 E. Jacobsohn, *Flimmeritis: Was jeder vom Kino wissen muss*, Berlin: Verlag der Lichtbild-bühne, 1919, p. 5.
2 M. Hansen, 'Early Cinema: Whose Public Sphere?', in T. Elsaesser (ed.) *Early Cinema: Space, Frame, Narrative*, London: BFI Publishing, 1990, pp. 228–46.
3 In considering these discussions of female cinemagoing, I am fully aware that I am not reconstructing the real historical audience but the various discourses that surround it. The chapter is thus more about how the audience was perceived than about how it actually acted. I rely here on the work done by Pearson and Uricchio regarding nickelodeon audiences in the United States: W. Uricchio and R. Pearson, 'Constructing the Audience: Competing Discourses of Morality and Rationalization During the Nickelodeon Period', *Iris* 17, 1994, 43–54. See also W. Uricchio and R. Pearson, '"The Formative and Impressionable Stage": Discursive Constructions of the Nickelodeon's Child Audience', in M. Stokes and R. Maltby (eds) *American Movie Audiences: From the Turn of the Century to the Early Sound Era*, London: BFI, 1999, pp. 64–75. However I will try to find traces of the actual historical audience, especially in the fan and fashion magazines.
4 E. Altenloh, 'A sociology of the cinema and the audience', *Screen*, 42.3, 2001, 249–93. Original: E. Altenloh, *Zur Soziologie des Kinos: Die Kino-Unternehmung und die sozialen Schichten ihrer Besucher*, Jena: Eugen Diederichs, 1914.
5 F. Kessler and E. Warth, 'Early Cinema and its Audiences', in T. Bergfelder, E. Carter and D. Görktürk (eds) *The German Cinema Book*, London: BFI, 2002, p. 124. Although the exact social composition of the first German movie audiences is still a much debated topic among scholars, most of them agree that the audience was above all hetero-geneous, that it contained workers as well as people from the middle classes, grown-ups and children, men and, of course, women. Cf. M. Loiperdinger, 'The Kaiser's Cinema: An Archeology of Attitudes and Audiences', in T. Elsaesser (ed.) *A Second Life: German Cinema's First Decades*, Amsterdam: Amsterdam University Press, 1996, pp. 41–50.
6 E. Altenloh, 'A sociology of the cinema and the audience', p. 275.
7 Ibid., p. 283.
8 Ibid., p. 257.
9 Ibid., p. 259.
10 Previously, the only unchaperoned women who walked the streets were prostitutes, while bourgeois women had been almost completely banned from the streets. Guide-books for young ladies published around 1900 still claimed that a girl could go any-where, but only on her father's arm.
11 V. Noack, *Der Kino: Etwas über sein Wesen und seine Bedeutung*, Leipzig: Dietrich, 1913, p. 3.
12 'Ein Schlag gegen die Kinotheater Wiens und Niederösterreich', *Der Kinematograph*, Nr. 264, 17.1.1912.
13 Ibid.
14 C. Moreck, *Sittengeschichte des Kinos*, Dresden: Paul Aretz, 1926, p. 212.
15 S. Hake, *The Cinema's Third Machine: Writings on Film in Germany, 1907–1933*, Lincoln, NA: University of Nebraska Press, 1993, p. 52.
16 Cf. for example A. Hellwig, 'Über die schädliche Suggestionskraft kinematographischer Vorführungen', in *Ärztliche Sachverständigen-Zeitung*, 20.6, 15.3.1914, 119–24. R. Gaupp, 'Die Gefahren des Kinos', in *Süddeutsche Monatshefte*, 9.2, Juli 1912, 363–66.

17 S. Curtis, 'The Taste of a Nation: Training the Senses and Sensibility of Cinema Audiences in Imperial Germany', *Film History*, 6, 1994, 445–69.

18 H. Schlüpmann, *Unheimlichkeit des Blicks: Das Drama des frühen deutschen Kinos*, Basel, Frankfurt a.M.: Stroemfeld / Roter Stern, 1990.

19 M. Grempe, 'Gegen die Frauenverblödung im Kino', in *Die Gleichheit*, 5, 1913, 70–72.

20 Hake, *The Cinema's Third Machine*, p. 53.

21 E. Lorenzen, 'Kinotypen', in *Bild & Film*, 2/8, 1912/1913, 188.

22 A. Knust, 'Ein Kinobrief aus der Provinzstadt', in *Lichtbild-Bühne*, 8(5) 1, 18.12.1915, 27.

23 R. Langer, 'Kinotypen. Plauderei aus den Lichtspielhäusern des Berliner Westens', in *Illustrierte Kino-Woche*, 18, 1914, 206.

24 'Pro und Contra den Kino', in *Der Kinematograph*, 282, 22.5.1912.

25 S. Harper and V. Porter, 'Moved to Tears: Weeping in the Cinema of Postwar Britain', *Screen* 37(2) 1996, 157.

26 A.M., 'Das neue Nollendorf-Theater in Berlin', in *Lichtbild-Bühne*, 5(13), 29.3.1913, 36. Judging from the number of copies sold *Sinful Love* (*Sündige Liebe*, D 1911, Deutsche Bioscop) had been one of the most successful films in 1911 in Germany.

27 For example G. Le Bon, *La psychologie des foules*, 1895; English translation: *The Crowd: A Study of the Popular Mind*, London: T. Fisher Unwin, 1896.

28 P. Schmidt, 'Irma das Kinogirl', in *Illustrierte Kino-Woche*, 3, 1913.

29 R. Wilke, 'Meine Frau im Film–Theater', in *Lichtbild-Bühne*, 8.52, 24.12.1915, 34–36.

30 S. Stamp, *Movie-Struck Girls: Women and the Motion Picture Culture after the Nickelodeon*, Princeton, NJ: Princeton University Press, 2000, p. 196.

31 Ibid., p. 197.

32 A. Walter, 'Erziehet die Kinobesucher', in *Lichtbild-Bühne*, 8.45, 6.11.1915, 46–48.

33 H. Schlüpmann, 'Die Erziehung des Publikums. Auch eine Vorgeschichte des Weimarer Kinos', in F. Kessler, S. Lenk and M. Loiperdinger (eds) *KINtop 5: Aufführungsgeschichten*, Frankfurt a.M., Basel: Stroemfeld / Roter Stern, 1996, 141–44.

34 C. Russell and L. Pelletier, ' "Ladies Please Remove Your Hats": Fashion, Moving Pictures and Gender Politics of the Public Sphere 1907–11', in *Living Pictures,* 2(1), 2003, 75.

35 [Jacobsohn], Egon, 'Ich weiß alles. Briefe vom Redaktionstisch', in *Illustrierte Filmwoche* [former *Illustrierte Kino-Woche*], 4, 1918.

36 'Sprechsaal', in *Illustrierte Kino-Woche*, 13, 29.8.1913. It seems as if the editorial office of the *Illustrierte Kino-Woche* had received so many film scripts by film enthusiasts that they wondered that 'the psychiatrists had not reported on this script writer craze', that 'awful epidemic'; 'Bericht über die 1. Versammlung der Vereinigung der Filmfreunde "Kurbelkasten"', in *Illustrierte Filmwoche*, 12/13, 1918.

37 *Elegante Welt* for example illustrated an article about vanity units with pictures of actress Maria Orska (20, 27.9.1916), and an article on billiards was illustrated with photos of Gunnar Tolnaes (18, 29.8.1917).

38 L. Kainer, 'Eleganz im Film', in *Die Dame*, 4, 15.11.1915, 1–5.

39 *Die Dame*, 9, February 1917.

40 'Henny Porten im Sport', in *Illustrierte Kino-Woche*, 9, 27.7.1913; 'Wie sich Kinokünstler erholen', in *Illustrierte Filmwoche*, 32/33, 1917.

41 See for example 'Bei Asta', in *Illustrierte Filmwoche*, 25, 1918 and 'Besuch bei Erna Morena', *Elegante Welt*, 16, 2.8.1916.

42 'Fern Andra auf der Rennbahn', in *Die Dame*, 19, 1.6.1916.

43 See for example *Illustrierte Kino-Woche*, 24, 1913; and *Elegante Welt*, 27, 2.7.1913 and numerous others.

44 *Elegante Welt*, 31, 5.8.1914.

45 R. Boelke, 'Die Verkäuferin', in *Illustrierte Filmwoche*, 34, 1918. See also 'Die kleine Maud. Aus dem Leben einer Filmenthusiastin', in *Illustrierte Kino-Woche*, 27, 1914, 'Plauderei. Der Herr Fuchs', in *Illustrierte Filmwoche*, 38, 1917 and 'Die kleine Filmschauspielerin', in *Illustrierte Filmwoche*, 45/46, 1917.

46 *Elegante Welt*, 22, 24.10.1917.

47 Schlüpmann, op. cit., p. 136.

48 J. Aubinger, 'Münchner Brief', in *Der Kinematograph,* 435, 28.4.1915.

49 A report from Munich in 1915 suggested that only one in ten visitors to a cinema was male. 'Zwölftausend Kinobesucher mehr', in *Lichtbild-Bühne,* 8.9, 27.2.1915, 9.

50 Cf. for example, P.M Grempe, 'Das Kino als Kulturbedürfnis', in *Der Kinematograph,* 453, 1.9.1915; 'Zwölftausend Kinobesucher mehr', in *Lichtbild-Bühne,* 8.9, 27.2.1915, 9.

51 H. Blaschitz, 'Jetzt – und ins Kino!?', in *Bild & Film,* 4/7, 8, 144.

52 N. Lutze, 'Die mondäne Frau im Lichtspielhause. Eine Betrachtung von Nanny Lutze', in *Der Deutsche Kaiser im Film,* Berlin: Verlag Paul Kleebinder G.m.b.H., 1912, p. 38.

10

AFGRUNDEN IN GERMANY

Monopolfilm, cinemagoing and the emergence of the film star Asta Nielsen, 1910–11

Martin Loiperdinger

The German film market in crisis

In 1908, the young German film industry was caught in its first capitalist crisis of overproduction. To get rid of superfluous films, producers and distributors lowered their prices and suffered losses on the capital invested in the production and purchase of films. In order to reverse the spiral of price dumping, the industry had to find a method of increasing and stabilizing prices. Since the abundance of existing short films did not offer any leverage, product innovations in film form were necessary. With every short film functioning merely as one item in a programme, film prices were calculated according to their length. Suppliers had to find a 'way out' of the short film programme, which, while popular with the public, was unprofitable. As Corinna Müller has demonstrated in detail, three innovations ultimately resolved the first crisis in the German film market. First, the expansion of the film length to at least two acts (two film reels) meant that a longer film replaced one or more short films in the programme. Second, the exclusive *Monopolfilm* distribution system guaranteed cinema owners that no other cinema in town could show the same film in the same week. The final innovation was the establishment of film stars as the deciding element in the public's choosing which films they went to see.[1]

Together, these three innovations accounted for the internal media upheaval in the German cinema industry's move from short film programmes to what became the standard programme format in Germany from the late 1910s to the 1970s, in which a long feature film was supported by a programme of advertising films, a newsreel and a *Kulturfilm* (an educational or travelogue one-reeler which German exhibitors usually programmed in order to have admission taxes reduced). The first steps in initiating this change were taken by a few managers of cinema chains in Denmark, script writer Urban Gad and actress Asta Nielsen, German film distributor Ludwig Gottschalk, and by many cinema owners and millions of patrons who paid to see *Afgrunden* (*The Abyss*, in German *Abgründe*, 1910).[2]

Afgrunden – Urban Gad and Asta Nielsen

The fundamental transformation of film as a commodity began in the Danish cinema sector in 1910. In that year, Fotorama, Denmark's largest film distribution company, which controlled a series of cinemas but was only of minor importance as a film producer, issued an unusually long feature film (706 metres) with the title *Den Hvide Slavehandel* (*The White Slave Trade,* in German *Die weiße Sklavin,* 1910). A film reel of about 300 metres normally constituted the upper limit for a short film included in programmes consisting of 'numbers'. With few exceptions, such as Passion Play films, a film 'number' seldom lasted longer than 12 minutes. At 35 minutes, *Den Hvide Slavehandel* was nearly three times as long. The topic of white slavery offered prospects of risqué scenes but, at the same time, it could be presented as morally acceptable, because an international treaty to combat white slavery had been signed at the beginning of May 1910. Fotorama's long production was evidently a great success with audiences: the police had to control the crowds queuing to see it in front of Fotorama's cinematograph theatre in Aarhus.[3]

Peter Elfelt, the owner of the Copenhagen cinematograph theatre Kinografen, produced another long feature film, *En Rekrut Fra 64* (*A Recruit of 1864,* 1910). At 970 metres long, the film dealt with the Danish-Prussian War, and began its run in Kinografen on 1 August 1910. It was directed by Urban Gad, artistic advisor and stage designer of Copenhagen's New Theatre, whose ensemble was not performing

FIGURE 10.1 Still from *Afgrunden*: Asta Nielsen performing the gaucho dance.

in the spring of 1910 because the theatre was occupied by a guest performance of Leo Fall's profitable operetta *The Dollar Princess*. Gad wrote the script of the third long feature film, the erotic melodrama *Afgrunden*, for his ensemble colleague Asta Nielsen. Anticipating a 'better audience', because it featured actors from reputable Danish theatres and was shown in luxurious cinemas, the film was a 'sex and crime' drama intended exclusively for adults. Asta Nielsen played Magda Vang, a young piano teacher who becomes engaged to an engineer, whose father is a pastor. When a travelling variety troupe comes to town, she succumbs to the call of freedom and adventure. She runs off with an actor, the variety theatre's 'cowboy', and becomes a member of the troupe. She performs a highly erotic variety number with her lover, the so-called 'gaucho dance'. She ropes the cowboy with her lasso and ties him up, to make him suffer tantalizing tortures. In a long, tight leather skirt, she dances around him, sways her hips, and rubs her body against his, while he has to remain still and cannot touch her. The men in the audience experienced the same torture: heightened by the actions on the screen, their desires were satisfied only through voyeurism. For their part, the women in the audience could identify with the lascivious seductress tempting the cowboy. Because of a conflict in the ensemble, the two lovers are forced to leave the travelling variety theatre troupe. Magda then earns money by playing the piano in cafes, while the former cowboy drinks away her money. During an argument he attacks her, and she stabs him in self-defence. The gaucho dance was banned in several countries, including Sweden, where the scene was saved in the censor's archives, and is, ironically, today the best-preserved scene from the film.[4]

Urban Gad was a friend of Hjalmar Davidsen, the owner of the Kosmorama in Copenhagen, which was one of a chain of over 20 cinemas in Danish provincial towns. Counting on Nielsen's natural and scandalously passionate acting style to appeal to the public, Davidsen invested 8,000 Kroner for a week's shooting in June of 1910, of which Nielsen was paid 200 Kroner for her performance.[5] At 850 metres (a running time of almost 47 minutes at 16 fps), *Afgrunden* was considerably longer than *Den Hvide Slavehandel*, and it dominated the film programme in his elegant capital city cinema, with a long run of several weeks at the Kosmorama.[6] This success did not, however, bring Nielsen any further engagement with the Danish film industry, not even with Davidsen himself, who earned close to 25,000 Kroner with the screenings of *Afgrunden* at his Kosmorama.[7]

In *Afgrunden*'s audience at the Kosmorama was the German film distributor Ludwig Gottschalk, owner of the Düsseldorfer Film–Manufaktur. Fifteen years later, he described his decision to go to Copenhagen:

> The unbelievable happened one day in the year 1910, incomprehensible for my distributor colleagues: I packed my bags and went to Denmark to have a look at films to buy for Germany. People thought I was crazy. The world belongs to the daring, I thought, and so I found myself sitting one evening in the Biograph Theatre [corr.: the Kosmorama] in Copenhagen before a three-act film of about 900 metres: *Afgrunden*, directed by Urban Gad and starring Asta Nielsen in the leading role. I was enthralled with the film and with Asta Nielsen.[8]

During the by then two-year-long crisis of the German film market, distributors had lost most of their creditworthiness through the continued reduction in rental prices. Facing the threat of bankruptcy, they were forced to think about a remedy. As early as Easter 1909, along with the usual short film programmes, distributors began to rent single films separately, offering them to larger cinemas as a special attraction. Initially, there was a shortage of suitable films, until the Danish company Nordisk plagiarized Fotorama with their own *Den Hvide Slavehandel* and sold copies of the 603-metre remake through their international representatives. From October 1910, several copies of this hit were in circulation on the German market. Cinema owners offered *Die weiße Sklavin* for sale or rental after they had used it.[9]

In this market situation, film distributors could gain an enormous advantage from obtaining exclusive rights to individual long films. Gottschalk's search for 'films to buy for Germany' did not involve unrestricted purchase of film copies on the open market, but buying an exhibition monopoly for the German market, and he arrived in Copenhagen just in time to secure the exclusive rights to *Afgrunden*: Davidsen had already sold other rights to the Skandinavisk-Russisk Handelshus company, and Nordisk tried to outbid Gottschalk, but the Düsseldorf film distributor eventually secured the exclusive German rights he sought.[10]

Abgründe in Germany – Ludwig Gottschalk's first *Monopolfilm* campaign

At the end of 1910, Gottschalk offered *Afgrunden* (under the title *Abgründe*) on the German market as a *Monopolfilm*, along with a much larger advertising campaign than had ever been seen before for a single film. He ran numerous full-page advertisements in the trade papers – a course of action that was totally unprecedented. Beginning on 16 November 1910, he placed a full-page advertisement in the trade journal *Der Kinematograph* almost every week for two months, offering German cinema owners the opportunity to secure the unrivalled premiere rights to this hit in their town for up to ten weeks.[11] He told cinema owners:

> In Copenhagen I succeeded in making a new conquest for the German theatre that will eclipse everything experienced up to now, will fill theatres for weeks, will be covered widely by the daily press, and will garner completely new audiences – a new audience for cinematography.[12]

Gottschalk based these promises on the claim that *Afgrunden* had been screened over 700 times in a Copenhagen cinematograph theatre during a run of eight weeks. His offer of exclusive distribution of the Danish 'theatre drama' was novel:

> With a great sum of money, I have acquired this hit *exclusively for Germany* and will offer the film on the market as early as next week. I will offer it for rental every week from the first to the tenth week and am prepared to grant premiere rights for individual towns for the first weeks.[13]

Having acquired the monopoly for all of Germany, Gottschalk could give the premiere rights to a cinema owner for a town, and the exhibitor could be certain that no other cinema in that area could screen the film. This new form of distribution was given the name *Monopolfilm*.[14]

On 30 November 1910, Gottschalk announced in his full-page advertisement: 'Only one opinion is heard about the theatre drama *Abgründe*: Realistic! Superbly thrilling! Make sure you obtain this hit. It will win you immeasurable success and sold-out houses for weeks.'[15] On 7 December 1910 the advertisement read: 'No cinematographic work has ever filled the box offices for such a long time.'[16] Three

FIGURE 10.2 Advertising for *Abgründe*.

weeks later, on 28 December 1910, Gottschalk ran New Year's wishes for 1911 and recommended booking *Abgründe*.[17] On 11 January 1911, Gottschalk promoted the movie 'as the greatest money maker for cinematograph theatres'.[18] A week later, he informed cinema owners that his offer was running out: 'On 28 January the last copies of the sensational film *Abgründe* ... will be on the market.'[19]

> *Abgründe* has been shown daily to sold-out houses: in Düsseldorf for 6 weeks, in Copenhagen for 8 weeks, in Berlin for 7 weeks, in the other large German cities for between 2 and 4 weeks without interruption with literally enormous box-office takes never seen before.[20]

Gottschalk supported his claims with quotations from letters sent to him by cinema owners. On 26 November 1910 the film had its German premiere at the Palast-Theater in Düsseldorf. On 7 December 1910 Gottschalk's advert declared:

> 'The Palast-Theater Düsseldorf has been stormed every evening for the last 14 days since *Abgründe* has been playing, and this theatre has requested to be allowed to keep this sensational film another 8 days. This is a record!'[21]

A three-week run for a single film was unique at the end of 1910.[22] On 28 December 1910, Gottschalk quoted the manager of the newly opened Theater des Weddings: *Abgründe*, which had been mentioned only briefly in the review of the new, 1,000-seat theatre's opening, had been sold out every evening as late as the third week.[23] Two weeks later, Gottschalk quoted a letter from the Palast-Theater in Breslau:

> *Abgründe* has created great excitement here, and, I think, a lot of people now realize that cinematography really is art. The demand for *Abgründe* is still great, and the box-office take has been the largest of the season up to now. I request that you inform me when I can have *Abgründe* again.[24]

Gottschalk's innovation promised enormous box-office revenues by combining the exclusivity of a long feature film with long running times. His success proved his case: the new *Monopolfilm* form of distribution allowed cinema owners to raise admission prices and created leeway for the distributor to raise rental prices.[25] Ultimately, however, the decisive factor in the business success of Gottschalk's *Monopolfilm* speculation was the audience. Overcoming the crisis in the film sector depended on building up a new habit of cinemagoing with an innovative product for the public. Gottschalk lauded the movie as the magnet that attracted patrons to cinemas. He made its impact on audiences appear as stimulus and response: as if patrons charged through the open theatre doors as soon as the billboards and newspapers announced the film title. Were Gottschalk's assertions only advertising claims, just rhetoric without much foundation, or did his advertising strategy point to real facts and figures?

An attraction for middle-class audiences

It seems that *Abgründe* actually did create a sensation among German audiences, at least in large cities. As is usually the case in research on historical audiences of early cinema, there are no direct sources produced by spectators themselves, such as diaries, autobiographies or readers' letters to the editor. As well as providing basic information on where and when a certain film was to be screened, the terms in which programme advertisements in daily newspapers addressed local audiences may indicate why the cinema owner who paid for the advertisement thought that the film would attract audiences.

Less than two weeks after the Düsseldorf premiere, Gottschalk circulated an advertisement in *Der Kinematograph* claiming that 'never has a theatre been attended by such a large number of the best viewers, as is now the case with the screenings of *Abgründe*'.[26] The expression 'large number of the best viewers' sounds very clichéd, but in fact refers to a surprising phenomenon described in some detail by the trade paper. After declaring that, when it came to matters of 'theatre, art and literature, Düsseldorf's "high society" conforms to the authoritative artistic community', it reported:

> Up to now, it was assumed that all film dramas were 'kitschte'. But *Abgründe*, showing at the Palast-Theater, has all of a sudden made artists enthusiastic fans of cinema pantomime. Everybody liked everything about this story and not least the gaucho dance, which in my opinion is inauthentic and – to put it mildly – indiscrete. Now my opinion, of course, does not count. The artistic community is very enthusiastic about *Abgründe*. I have met almost no artist or actor there who had not seen *Abgründe* for the third, fourth, fifth or even the eighth time … At any rate, *Abgründe* has contributed a great deal, if not the most, to turning the conversation in society more than usual to film theatre and to leading people you would never expect to the cinema.[27]

The second claim in Gottschalk's advertisements was that no single film had such a long running time, and that no single film generated such box-office revenue during that time. The letter from Breslau that Gottschalk used as an advert provided evidence of extended running times, and can be verified by facts and figures. Breslau (today Wroclaw, Poland) was then the German centre of the Silesian border region and among the ten most important cities in the German Reich. The film was shown at the 450-seat Palast-Theater, in the evening programme from 20 till 30 December 1910.[28] The cinema owner, Franz Thiemer, tried to obtain another copy of the film for further showings as soon as possible. Only one week after Gottschalk had published Thiemer's letter, *Abgründe* was back at the Palast-Theater for another two weeks, but now it was screened twice an evening.[29] Since the last advertisement claimed that *Abgründe* had been shown 86 times in Breslau, it seems that the print was also shown in Thiemer's second cinema, the Monopol-Theater.[30] No other single film more than one reel long had been shown so many times before in Breslau and the frequency of screenings is a clear indication of the scale of audience attendance in Breslau.[31]

The most convincing evidence for *Abgründe* being the starting point for the media upheaval that marks the transition from the short film programme to the long feature film format may be found in the Hamburg area. In 1910, Hamburg had a population of more than 900,000, and the city merged into an industrial agglomeration of 1.2 million inhabitants with nearby towns such as Altona and Eimsbüttel.[32] The local newspaper *Hamburger Fremden-Blatt* contains reports and advertisements for entertainment enterprises, providing evidence of a rich and lively public life, but historical research on local cinemas has to rely completely upon the advertisements of the cinema owners, since the reviews covered theatre, concerts, variety theatres, the wax museum, and stereo panoramas, but not cinemas. In this instance, at least, Gottschalk was mistaken when he expected that *Abgründe* would be covered widely by the daily press.[33] It took two more years before Hamburg newspapers were willing to report on films premiering in the city.[34]

According to cinema advertisements in the *Hamburger Fremden-Blatt*, *Abgründe* screenings started in Altona, Eimsbüttel and Hamburg in the middle of January 1911. By mid-March, it had been shown 223 times. From 14 to 20 January it screened three times a day at two cinemas, the Belle Alliance-Theater in Eimsbüttel and the Helios-Theater in Altona. From 21 to 29 January it screened at the Palast-Theater and in the Victoria-Theater in Hamburg. All four cinemas were owned by James Henschel, the uncrowned king of cinemas in the Hamburg area, who was able to screen *Abgründe* 90 times within a fortnight. Over the next week it screened between three and five times a day at Henschel's Waterloo-Theater and the Hammonia-Theater, owned by Hermann Kasper. When it was banned for a week by the Hamburg police in early March, the film returned to Henschel's two theatres in Eimsbüttel and Altona. When the ban was removed, it was shown twice a day from 11 to 17 March in the Elite-Theater (where it returned for another week-long run from 27 May to 2 June).[35] Overall, the film ran for seven weeks between mid-January and mid-March. No cinema programme or single film had previously been offered to patrons in Hamburg for such a long period: *Abgründe* marked the starting point of a new era in film exhibition.

Abgründe's position in German cinema history is even more striking when we examine the seating capacity that Hamburg cinema owners reserved for it. Combining the adverts for *Abgründe* in *Hamburger Fremden-Blatt* with local research by Michael Töteberg as presented in Table 10.1, we can establish that 184,182 seats were available for viewing the film from mid-January to mid-March.[36] Since children under 16 (who comprised approximately 35 per cent of the population), were not permitted at these screenings, and since we can assume that the additional 5 per cent of the population over 65 probably did not attend, the film's potential audience in the Hamburg area was around 700,000 people, meaning that a seat was available for every 3.8 potential patrons. Pushing a single film so extensively into the channels of cinema exhibition was without precedent. This practice was certainly unique up to that time, not only in the cinema business, but in the whole entertainment sector of the Hamburg area.

TABLE 10.1 Screenings of *Abgründe* in Hamburg, January to March 1911

Theatre	Seats	Period	Days per theatre	Screenings per day	Total screenings	Seating for Abgründe
Belle Alliance-Theater	1,191	14–20 Jan	7	3	21	25,011
Helios-Theater	500	14–20 Jan	7	3	21	10,500
Palast-Theater	1,600	21–29 Jan	8	3	24	38,400
Viktoria-Theater	250	21–29 Jan	8	3	24	6,000
Waterloo-Theater	900	30 Jan–3 Feb	5	4	20	18,000
Hammonia-Theater	535	30 Jan–3 Feb	5	3	15	8,025
Hammonia-Theater	535	4–10 Feb	7	3	21	11,235
Waterloo-Theater	900	25 Feb–3 Mar	7	5	35	31,500
Belle Alliance-Theater	1,191	4–10 Mar	7	3	21	25,011
Helios-Theater	500	4–10 Mar	7	3	21	10,500
TOTAL			**68**		**223**	**184,182**

Abgründe was a success from the beginning. At a time when many theatres changed their short film programmes twice a week, the comparatively long periods of continuous screenings and the re-release after several weeks or even months provide evidence of this. Success in this case meant that local audiences continued to attend in considerable numbers even after several weeks of running time. It seems that the phrase used in many advertisements, 'due to great demand', was not simply an empty expression in this case. What information in the cinema advertisements might have made the film attractive for the public?

Compared to short film programmes, the advertisements offered an ambiguous mixture. *Abgründe* was announced as a 'sensational theatre drama', which at the time implied sophistication (theatre) and sex and crime (sensation). This second aspect was emphasized by the exclusion of children, even when accompanied by adults. Although this eliminated an important segment of the audience for short film programmes, the exclusion of children during the evening screenings may have attracted patrons accustomed to going to theatre performances. As early as the beginning of 1907, cinemas in big city centres had tried to compete with theatres in the exterior and interior appearance of their buildings, but these *Kinematographen-Theater* showed more or less the same short film programmes as the cheap storefront theatres. With a 'theatre drama' played by theatre actors from Copenhagen, they now had something different in both content and style from what had previously been seen on screen to attract the middle-class audiences who read advertisements in local newspapers and may also have responded to the information that some theatres' directors of music had arranged a score especially for *Abgründe*. When Franz Thiemer told Gottschalk, 'I think that a lot of people now realize that cinematography is really art', he was talking about middle-class audiences who had never before considered cinemagoing worthwhile entertainment. Gottschalk quoted this sentence to assure cinema owners that the film promised to attract the so-called 'better' people as well as adult viewers already accustomed to cinemagoing.

Abgründe and the Asta Nielsen *Monopolfilm* series: emergence and making of the film star

Following *Abgründe*'s success, the Cologne film dealer and cinema owner Christoph Mülleneisen, Sr. devised a business model to introduce the star system to the German film industry 'which is still impressive in its professionalism, imaginative elegance and daring even today'.[37] While the *Monopolfilm* distribution system offered exclusive exhibition rights, Mülleneisen extended the monopoly on the production side, by exclusively hiring a renowned actress for a number of long feature films, at a time when German production companies usually hired actresses and actors by the month. On 27 May 1911, Mülleneisen, PAGU (Projektions-AG Union) and Österreichisch-Ungarische Kino-Industrie GmbH founded the Internationale Films-Vertrieb-Gesellschaft (IFVG) with headquarters in Vienna. The new company selected as its manager Paul Davidson, who headed PAGU, and contracted Asta Nielsen and Urban Gad to work exclusively for it, producing eight long feature films a year until 1914. Nielsen was to star in each film, and Gad was to write and direct.[38] Six weeks later, the new company announced that it intended to invest 1,400,000 Marks in the production of Nielsen's *Monopolfilms* in the next six months. 'That means being gutsy!', chimed the advertisement. Convinced that her presence in the *Monopolfilm* series guaranteed that cinema owners would book them in advance, even before they were produced, the producers made Nielsen the star of the enterprise. From August 1911, she was given the label 'the Duse of film art', and placed on a level with the world-famous Italian star of the stage.

Gottschalk's successful experiment was a prelude to the introduction of the star system to the German film market. During the film's first months of exhibition Nielsen emerged as a star: in his initial advertising campaign, Gottschalk emphasized the writer by calling *Abgründe* a 'theatre drama in two acts by Urban Gad'.[39] Only in his last advert did he mention the names of the roles portrayed 'by famous members from Copenhagen theatres', with Nielsen first on the list.[40] After the film's box-office success, advertisements for her films *Heißes Blut* (*Burning Blood*, 1911) and *Nachtfalter* (*The Moth*, 1911), called her 'Schöpferin der *Abgründe*' ('the creator of *Abgründe*').[41]

As a prelude to the introduction of the star system, *Abgründe* strikingly reveals some essential aspects of cinema's modernity. Although the distributor originally promoted the writer-director, the audience pushed the actress into the focus of public attention. Asta Nielsen was the first star in Germany's entertainment business whose emergence took place completely through technological reproduction. Audiences established personal and emotional relations with a hitherto unknown foreign actress without having seen her in person. Compared to live acts on stage, Nielsen's performance was much more widely disseminated and available; in Hamburg one third of the adult population had seats available to them.

A few years before the outbreak of the First World War, modernity in the entertainment sector was characterized by the merger of powerful capitalist and democratic forces into what was later called 'consumerism'. The upheaval in cinema programming which *Abgründe* initiated in Germany provides an instructive illustration of this process. The *emergence* of the film star Asta Nielsen took place because of the crowds who occupied the seats that were offered by cinema owners, but as soon as it had become obvious that audience response mainly focused on Nielsen, the *making* of the film star Asta Nielsen was quickly decided on by a group of businessmen, who did not hesitate to invest an enormous amount of money in order to profit from her future career. Audience attendance at *Abgründe* demonstrated that it was possible to overcome the enduring crisis in the German film market. The introduction of the star system proved to be an outstanding means of again earning a profit from the production, distribution and exhibition of films.

Notes

1 C. Müller, *Frühe deutsche Kinematographie. Formale, wirtschaftliche und kulturelle Entwicklungen*, Stuttgart, Weimar: J.B. Metzler, 1994.
2 The film's American release title was *The Woman Always Pays*.
3 M. Engberg, 'The erotic melodrama in Danish silent films 1910–18', *Film History* 5, 1, 1993, 65.
4 J. Olsson, ' "Dear Miss Gagner!" – a star and her methods', in T. Soila (ed.) *Stellar Encounters: Stardom in Popular European Cinema*, New Barnet, Herts: John Libbey, 2009, pp. 217–29. *Afgrunden* is available on a DVD with the four early Danish films starring Asta Nielsen. *Asta Nielsen. Danish Silent Classics Nr. 15.* DVD, Danish Film Institute, 2005. This version of *Afgrunden* runs for 35 minutes.
5 A. Nielsen, 'Mein Weg im Film. 2. Mein erster Film', *B. Z. am Mittag*, 24 Sept. 1928, reprinted in R. Seydel and A. Hagedorff, *Asta Nielsen. Ihr Leben in Fotodokumenten, Selbstzeugnissen und zeitgenössischen Betrachtungen*, Berlin: Henschelverlag, 1984, p. 38.
6 Advertisements by Düsseldorfer Film-Manufaktur speak of a run of eight weeks, however, already five weeks after the Copenhagen premiere. Cf. *Der Kinematograph* 203, 16 Nov. 1910. Title page reprinted in Müller, *Frühe deutsche Kinematographie,*p. 116.
7 M. Engberg, *Filmstjernen Asta Nielsen*, Aarhus: Klim, 1999, p. 53. I wish to thank Patrick Vonderau for the explanation of the Danish text.
8 L. Gottschalk, '15 Jahre "Monopolfilm"', *Lichtbild-Bühne* 18.237, 21 Nov. 1925, 14.
9 Müller, *Frühe deutsche Kinematographie.* pp. 110–15.
10 Engberg, *Filmstjernen Asta Nielsen*, pp. 53–54. In 1924, Maxim Galitzenstein, former marketing director of the Messter company, gave a completely different version of Gottschalk's purchase of the *Afgrunden* rights. According to him, Hjalmar Davidsen and Urban Gad had travelled to Berlin to sell *Afgrunden* and, because of the film's length, did not arouse much interest, so that Gottschalk was able to obtain it at a bargain. M. Galitzenstein to O. Messter, 24 November 1924, Bundesarchiv NL 275, Akte 355. See also Müller, *Frühe deutsche Kinematographie*, p. 306.
11 *Der Kinematograph* 204, 23 Nov. 1910.
12 Cover advertisement. *Der Kinematograph* 203, 16 Nov. 1910. Reprinted in Müller, *Frühe deutsche Kinematographie*, p. 116.
13 Müller, *Frühe deutsche Kinematographie.*
14 The German word 'Monopolfilm' has no direct English equivalent. Rachel Low talks about 'the system of exclusive hire', or 'exclusive films', or just 'exclusives'. R. Low, *The History of the British Film, 1906–1914*, London: George Allen & Unwin, 1949, pp. 44–49.
15 *Der Kinematograph* 205, 30 Nov. 1910.

16 *Der Kinematograph* 206, 7 Dec. 1910.
17 *Der Kinematograph* 209, 28 Dec. 1910.
18 *Der Kinematograph* 211, 11 Jan. 1911.
19 *Der Kinematograph* 212, 18 Jan. 1911.
20 Ibid.
21 *Der Kinematograph* 206, 7 Dec. 1910.
22 As seen in the adverts of the Düsseldorf Palast-Theater, the film ran for three weeks. Cf. the adverts in the *Düsseldorfer Neueste Nachrichten* 276, 26 Nov. 1910; 288, 10 Dec. 1911.
23 'Ein neuer Tempel der Kunst!' Review of the premiere, *Lichtbild-Bühne* 124, 10 Dec. 1910. The running time of *Abgründe* in the Theater des Weddings cannot be determined from adverts in the Berlin newspapers. Adverts could be found only in the *Berliner Morgenpost* 338, 10 Dec. 1910; 339, 11 Dec. 1910; 346, 18 Dec. 1910.
24 Letter from the Palast-Theater, Breslau, quoted in an advertisement in Ludwig Gottschalk's 'Düsseldorfer Film-Manufaktur', *Der Kinematograph* 211, 11 Jan. 1911.
25 Gottschalk himself is said to have earned 45,000 gold marks on the royalties of *Afgrunden* in Germany and elsewhere (according to his niece Mary Furstenberg). Cf. K. G. Jaeger, 'Die "Düsseldorfer Film-Manufaktur"', *Publizistik* 28.1, 1983, 93.
26 *Der Kinematograph* 206, 7 Dec. 1910.
27 Niko, 'Düsseldorf im Januar 1911', *Der Kinematograph* 213, 25 Jan. 1911.
28 Adverts of Palast-Theater. *Breslauer General-Anzeiger* 345, 18 Dec. 1910; 347, 20 Dec. 1910; 348, 21 Dec. 1910; 350, 23 Dec. 1910; 356, 30 Dec. 1910.
29 Adverts of Palast-Theater, *Breslauer General-Anzeiger* 17, 18 Jan. 1911; 18, 19 Jan. 1911; 22, 22 Jan. 1911.
30 Advert of Palast-Theater, *Breslauer General-Anzeiger* 31, 1 Feb. 1911.
31 M. Loiperdinger, '*Abgründe* – poczatek dlugometrazowych filmów fabularnych we Wroclawiu', in A. Debski and M. Zybura (eds) *Wroclaw bedzie miastem filmowym. Z dziejów kina we Wroclawiu*, Wroclaw: Wydawnictwo GAJT, 2008, pp. 177–86.
32 According to the census of December 1910, Hamburg had 931,035 inhabitants, Altona 172,628, and Eimsbüttel 116,400.
33 Gottschalk's first advert, *Der Kinematograph*, 16 Nov. 1910, quoted above.
34 C. Müller 'Kinoöffentlichkeit in Hamburg um 1913', in C. Müller and H. Segeberg (eds) *Kinoöffentlichkeit (1895–1920) – Entstehung, Etablierung, Differenzierung / Cinema's Public Sphere – Emergence, Settlement, Differentiation (1895–1929)*, Marburg: Schüren Verlag, 2008, pp. 105–25.
35 All information on the running times and programming of *Abgründe* in the Hamburg area are taken from cinema adverts in the *Hamburger Fremden-Blatt* which were usually published on Saturdays.
36 M. Töteberg, 'Neben dem Operetten-Theater und vis-à-vis Schauspielhaus. Eine Kino-Topographie von Hamburg 1896–1912', in Müller and Segeberg, *Kinoöffentlichkeit*, pp. 87–104. Figures for Helios-Theater and Victoria-Theater are taken from Film- und Fernsehmuseum Hamburg: www.filmmuseum-hamburg.de.
37 Müller, *Frühe deutsche Kinematographie*, p. 144.
38 Some sources suggest that Nielsen and Gad were contracted to produce ten films per year. Nothing is known of the whereabouts of this important contract. M. Loiperdinger, 'Der erste Filmstar im "Monopolfilm"-Verleih', in H. Schlüpmann et al. (eds.) *Unmögliche Liebe. Asta Nielsen, ihr Kino*, Vienna: Verlag Filmarchiv Austria, 2009, pp. 177–86.
39 *Der Kinematograph* 206, 7 Dec. 1910; 209, 28 Dec. 1910; 211, 11 Jan. 1911.
40 *Der Kinematograph* 212, 18 Jan. 1911.
41 *Der Kinematograph* 228, 10 May 1911.

11

'LITTLE ITALY ON THE BRINK'

The Italian diaspora and the distribution of war films in London, 1914–18

Pierluigi Ercole

On 26 April 1915, the Italian Prime Minister signed a treaty with Britain and the other Entente Powers in London. The agreement established that the *terre irredente* (unredeemed territories) of Trento, Istria, Fiume, Trieste and Dalmatia would be given to Italy at the end of the war on the condition that the latter was ready to join the conflict within one month. Until then, Italy had maintained a neutral position. The decision to join the Entente Powers came after a long period of intense political and social debate that had divided Italy between supporters of neutrality and interventionism. On 23 May 1915, nine months after Britain had declared war on Germany and Austria-Hungary, Italy entered the war on the British side. The following day, the *Manchester Guardian* newspaper reported:

> In the headquarters of the ice-cream business – shy streets that dip down from the tram-ridden Clerkenwell Road – one searched in vain yesterday for war fever. The grizzled men in shirt sleeves sitting out on the soiled doorsteps of dark little shops were as cool about the crisis as the stuff they sell. The notice of the 'dimostrazione patriocca' (sic) on the walls did not stir more than a leisurely Southern interest. How different from the ebullient excitement of our Stock Exchange. 'Little Italy' was sunning itself within its smeared and cosy limits, only mildly interested apparently in 'Italy on the brink', 'Italy's fateful hour', and other signs of agitation in those English newspapers which few Little Italians can read.[1]

Despite such an apparently lacklustre atmosphere within the Italian quarter itself, the newspaper reported that later that day, 'the legions of Clerkenwell marched west on Grosvenor Square and besieged the Italian Embassy ... The sun lit up prettily the soiled banners of the London Italian societies floating at the head of the procession, behind came a band in khaki and in the place of honour a Garibaldian veteran in his red shirt and kepi'. Unfortunately on this occasion 'the Embassy blinds were down' and the efforts of the Italians to rouse the Ambassador were in vain, 'so after the Italian national song and "God Save the King" the banners fluttered away again, and Little Italy, hot and happy, went home to tea'.

Three days later, on 27 May, a second and larger demonstration organized with the endorsement of the Italian Ambassador took place, and hundreds of Italians, cheered on by crowds of Londoners, marched from the Embankment through the capital's centre to Grosvenor Square. From the balcony of the Italian Embassy, the Ambassador addressed, in Italian, his 'brothers of Italy', and British MPs 'paid a hearty and sincere appreciation to Italy's action in joining [Britain] as an ally'.[2]

This chapter presents a case study of the involvement of Italian immigrants in the screening of Italian war films in London during the First World War. From before the May demonstrations, two Italian immigrants involved in the British film industry, doctor Mario Pettinati and chevalier Arrigo Bocchi, began to organize special events involving the capital's Italian community at which they screened topical Italian films depicting the operations of the nation's military. These events had two main goals: to reinforce a sense of national identity and belonging among the immigrant colony, and to show the British public the war effort of their Italian allies. Mirroring the nationalistic discourses that pervaded the London-based Italian press of the time, Pettinati and Bocchi's screenings sought to promote national values based on the history of Italy's political and geographical unification since 1861.[3] Central to my discussion is the role played by cinema in creating a sense of national belonging and identity among the members of the immigrant community. Before detailing cinema's cultural role it is, therefore, important to consider the history and cultural specificities of the Italian community in London. In addition, I want to contextualize the reception of Italian war films within the distribution of Italian productions in Britain.

The micro-historical focus of this chapter reaches out and intersects with a wide range of issues related to modernity, national identity and diasporic culture, as well as film distribution and reception. The study of the relationship between Italian immigrants and film culture in early twentieth-century London brings to the fore a variety of interrelated questions that are central to the debate about the social experience of moviegoing at the turn of the century. Together with processes of nation-state building, mobility and circulation of both commodities and populations were social, political and cultural phenomena that characterized modern society. Commenting on the periodization of modernity suggested by Stephen Kern in his book *The Culture of Time and Space*, Ben Singer suggests that the period between 1880 and the end of the First World War 'saw the most profound and striking explosion of industrialization, urbanization, migration, transportation, economic rationalization, bureaucratization, military mechanization, mass communication, mass amusement, and mass consumerism'.[4] As the Lumières' train was repeatedly arriving at its station in the 1890s, great waves of Italians were leaving the peninsula and thousands of them were settling in London's Little Italy. In 1905, the Italian film production industry began to emerge with the foundation of the Turinese company Carlo Rossi & C. (later Itala Films), whose 1907 film *Il Cane Geloso* (*The Jealous Dog*) was one of the first Italian moving pictures to be seen on British cinema screens. During the following seven years, Italian film production flourished, in a period now commonly referred to as the Golden Age of Italian cinema, and hundreds of films crossed the Channel to be screened in British cinemas. At the beginning of the twentieth century, then, London had become a prime site for the convergence of both Italian migrant people and moving images.

Italian films in Britain

Between 1907 and 1914 more than three thousand Italian films were made available to British cinema owners.[5] The significant presence of Italian productions in the British market during this period positioned Britain as one of the leading importers of Italian films. Production companies such as Cines, Ambrosio, Milano Films and Itala Films found across the Channel the right conditions, both in terms of distribution and critical reception, for widening the film market beyond national borders. London, as the central hub for film industry activity, played a significant role in the distribution of Italian productions both in Britain and beyond.

From 1907 to the First World War the majority of Italian films released in Britain were fiction titles, with non-fiction films constituting only about 15 per cent of the imports. Although some of the Italian non-fiction films released in Britain were industrial or educational films, the majority were topical shorts, in the form of scenic films and travelogues, creating an image of Italy for British audiences, whether they were native or immigrant. In 1911, for instance, over half the travelogues and scenic films showed Italian panoramas, cities and monuments.

Titles such as *Le Marine Del Lazio* (*Picturesque Sea View Near Rome,* 1910), *Napoli* (*Naples,* 1910), and *Inaugurazione Del Monumento A Vittorio Emanuele* II (*Unveiling of King Emanuel's Monument in Rome,* 1911) were part of an impressively detailed atlas of Italy in moving images that companies such as Cines, Latium, and Aquila were exporting to Britain. With the emergence in the seventeenth century of the Grand Tour as a 'new paradigm for travelling', Italy, home of classical traditions, became one of the most important destinations without which the education and knowledge of well-bred British travellers was not complete.[6] For such travellers, a journey through the most important Italian cities such as Venice, Florence, Rome and Naples provided a means of developing their education in art history and cultivating the 'principles of correct taste'.[7] The fascination for antiquities and picturesque views of Italian landscapes also created a tradition of visual representation that found in J.M.W. Turner its major exponent.[8]

Within this already established cultural context, Italian travelogues and scenic films were the modern expression of the tourist tradition. Although mediated by the camera, the scenery portraits offered to every class of audience the possibility of experiencing a Grand Tour of Italy and discovering its beauties. The artistic and natural richness of the country revealed itself as an inexhaustible source for the Italian film companies. To the British audiences, travelogues of Italy functioned as modern tourist postcards; to the Italian producers these films were the expression of national pride.

While non-fiction films portrayed the natural and artistic beauties of the peninsula, Italian-made dramas conveyed a stronger sense of Italian-ness. The efforts by Italian film companies to promote a pedagogical function for film, and the aspiration to be known abroad for their high quality and high-brow productions, were particularly represented in this genre of films. A significant number of the Italian dramas released in Britain made direct references to Italian culture and for the most part this was obvious from their titles, even in the Anglicized versions. In large part these films were biographical and historical dramas that depicted major cultural figures and epochs from Italy's past, from ancient Rome to the unification of the nation in 1861.[9]

In 1911 alone, 47 of the imported films were based on historical events and set in the peninsula, 12 of them in ancient Rome. The first film dealing with the Roman Empire to reach British screens was *Gli Ultimi Giorni Di Pompei* (*The Last Days of Pompeii*, 1908) which was released in Britain in February 1909. The vicissitudes of the Roman Empire were also re-enacted in *Nerone* (*Nero*, 1909), *Spartaco* (*Spartacus, the Last of the Gladiators*, 1909), *Martire Pompeiana* (*The Pompeian Martyr*, 1909), *Guilio Cesare* (*Julius Caesar*, 1909), *Principessa E Schiava* (*Princess and Slave*, 1909), *Amore Di Schiava* (*The Slave's Sacrifice*, 1910), *Messalina* (*The Love of an Empress*, 1910), *Bruto* (*Brutus*, 1910), and *Clio E Filete* (*Clios and Phyletes*, 1911), among others. One of the most celebrated examples of such films was *Quo Vadis?* (1913), a major international success and a highly influential production. Reviewing *Quo Vadis?* the British trade paper *The Bioscope* wrote:

> It is not often that the work of reviewing film plays allows one to indulge in unmitigated praise, to luxuriate in superlatives, because however excellent the creations of modern producers, and however steady their artistic development, it is seldom that any one of them so far overshoots the combined accomplish-ment of his fellows as to bring forth a single masterpiece which in itself marks a distinct epoch in the history of the cinematograph art. This, however, is what has been done by the maker, or makers, of *Quo Vadis?*.[10]

The more recent history of the unification of Italy was dealt with in *Il Piccolo Gar-ibaldino* (*The Garibaldinian Boy*, 1909), *Anita Garibaldi* (1910), *Erocio Pastorello* (*Heroic Shepherd Boy*, 1910), *Nozze D'oro* (*After Fifty Years*, 1911), *Goffredo Mameli* (1911), *Stirpe D'eroi* (*Mother of the Heroes*, 1911) and *Il Tamburino Sardo* (*Sardinian Drummer Boy*, 1911). The significant number of such pictures shown in British cinemas meant that, in purely quantitative terms, the films could play an important role in creating a cultural imagination of Italy. Visualizing crucial events of Italian history, the historical productions presented an aura of respectability to both Italian and British audiences. In her study on the cultural and ideological values of cinematic representations of Rome, Maria Wyke stresses that before the First World War in Italy:

> Film production was viewed as an instrument for the enhancement of the new nation's prestige both at home and abroad, and historical reconstructions of Italy's glorious past seemed highly appropriate vehicles for the acquisition

of that prestige both for the Italian nation and its film companies […]. Films set in Italy's Roman past were perceived and deployed as instruments particularly suited to the moral, civic, and patriotic improvement of their mass audiences. The Italian state born from unification in 1861 continued to view itself as the legitimate heir of its Roman past […]. Rome could therefore supply the Italian film companies with a repertoire of illustrious precursors through whom audiences could read their present as the crowning epoch of a long glorious and communal history.[11]

The production of Italian historical films has to be read within its own cultural context and in particular in conjunction with Italian imperialistic and colonial ambitions, which culminated in the invasion of Libya in 1911, the year of the fiftieth anniversary of Italian unification. The production of films dealing with the glorious past of Italy, and more specifically of the eternal city, mirrored the nationalist sentiments that pervaded not only Italian political and cultural life, but also the popular imagination of the Italian people before the Great War. This spirit of nationalistic pride pervaded the content of the London-based Italian newspaper *Londra-Roma*, which represented the Italian colony in London as being unified under a sense of belonging to a prestigious 'Madre Patria' (Mother Country). The Italian historical film productions available in considerable quantity to Italian immigrant audiences in London echoed visually the colony's patriotic discourses.

London's Italian colony

By 1915, there were approximately 11,000 Italian immigrants living in two main communities in London. The older and larger community, popularly known as Little Italy and situated in the Clerkenwell district of the capital, was formed at the turn of the nineteenth century. A second community established itself in Soho in London's West End in the 1890s. According to the Italian census, between 1871 and 1915 roughly 15 per cent of the total number of Italian immigrants who came to Britain originated from the south of Italy, with the majority coming from central and northern regions. By the turn of the century, however, the number of southern Italians arriving in Britain was increasing, and most of those who settled in the capital tended to live in Little Italy. A large proportion of London's Italian immigrant population were unskilled peasants, and many of Little Italy's residents took up work as itinerant street entertainers and hawkers, such as barrel-organ players and ice-cream sellers. Many of the primarily northern Italians residing in Soho, on the other hand, were connected with the catering trade, working as waiters, kitchen staff and employees of hotels and lodging-houses.[12] As historian Terri Colpi has identified, the Soho Italians 'rarely became organ-grinders and were not so much connected with the itinerant street selling of food' and 'were rather more sophisticated and better off' than the Clerkenwell colony.[13]

   There is little indication that Italian immigrants living in either of the two main colonies in London had achieved any significant level of assimilation into wider

London society by the early 1910s. The Italians were depicted as a contained and closed community, with their own religious customs and traditions brought with them from their villages of origin.[14] In addition, few Italians became naturalized British citizens and their isolation from the English community was often seen as self-imposed: according to the *Daily Graphic* in 1905, 'the inhabitants of Little Italy do not seek the companionship of their English neighbours, and they resent the intrusion of visitors'.[15] Accounts by British journalists and social commentators detailing the capital's Italian immigrants focused largely on Little Italy and tended to present the colony and its inhabitants as both a curiosity and a threat, emphasizing colourful details amidst a dark and dangerous landscape. Despite their obvious prejudice, these accounts certainly suggest the difficulties in the Italians' relationship with their host city. To outsiders looking in, the Italians were a largely illiterate underclass whose retention of their native culture and Italian nationality marked them out as both picturesque and alien, and at the same time provided 'evidence of [their] tenacious patriotism' to Italy.[16]

Italian patriotism and the screen

Apart from the two main London-based Italian communities, a number of more affluent Italians, including members of the artistic and intellectual elite, aristocrats and diplomats, were actively engaged in promoting both Anglo-Italian relations and national pride amongst the immigrant community. For instance, in 1914, a series of 'patriotic lectures' centred around Italy's unification were delivered by Tullio Sambucetti, Secretary of the Italian Chamber of Commerce in London. According to *Londra-Roma*, Sambucetti addressed the audience in simple and plain Italian in order to make his lesson accessible to the ordinary working-class immigrants. Sambucetti's paper made reference to Italy's glorious past and emphasized that the country's unification had given rise to a 'new spirit in Italy', which, in turn, would lead to a moral and economic transformation of the country.[17] Sambucetti's lecture series ran throughout the summer of 1914. Just weeks after Britain had declared war on Germany and months before Italy's intervention in the conflict, a section of London's Italian community began to mobilize support for the British cause. In late September, Sambucetti delivered a speech at a rally at the Queen's Hall organized to demonstrate Italy's solidarity with Britain.

As the war progressed and the moment of Italy's intervention drew nearer, there was an increasing programme of activities designed to galvanize patriotic sentiment amongst the immigrants and to showcase Italy's support for Britain, and moving images began to play a visible role in this series of events. In particular, two Italian immigrants, Arrigo Bocchi and Mario Pettinati, were responsible for organizing patriotic screenings involving the Italian community. Close friends of Marquis Guido Serra di Cassano, the well-known manager of the Italian Cines Film Company's offices in London, both Bocchi and Pettinati had become involved in the British film industry during the 1910s. Bocchi had, in fact, started out in London in the 1890s as a composer and musician and he became well-known not only within the immigrant community, but also in wider London society. In December 1913, Bocchi became the manager of the newly opened Philharmonic Hall, which was used as a venue primarily for musical concerts and later for the screening of films. Starting in 1914, the Hall was used as a cinema, hosting screenings of topical films for extended periods including Herbert G. Ponting's *With Captain Scott In The Antarctic* (1913), and a series of Kinemacolor films depicting R.G. Knowles' travels to Imperial India.[18] In 1916, Bocchi became involved in film production when he established the N.B. (Nash Bocchi) Films Company together with the British producer-director Percy Nash, before joining the Windsor Film Company, which was owned by Guido Serra, sometime before 1918.[19]

During the 1910s Mario Pettinati worked as a correspondent for the Italian newspaper *Gazzetta del Popolo*. According to the 1924 *Kinematograph Year Book* Pettinati had been connected with the cinema industry since 1906.[20] In 1913, he co-founded two companies in Britain: the Italian Press and General Agency was a press, publishing and advertising agency, and the London and Counties Film Bureau distributed mostly Italian produced films.[21]

In January 1915 Bocchi organized a *festa patriottica* (patriotic celebration) at the Philharmonic Hall, the proceeds of which were given to the *Società Nazionale Dante Alighieri*, a London-based Italian association that ran a school for children of the immigrant community. According to *Londra-Roma*, members of every class of the Italian colony 'from diplomatic and consular authorities to the most modest work-man' attended the event.[22] At the celebration the audience was entertained with comical sketches performed on the stage, while a programme of topical films depicting the Italian army and navy provided a more serious note to the patriotic proceedings. The pictures for which the audience showed the most appreciation were those featuring scenes of the *Alpini*, the mountain warfare soldiers of the Italian Army, transporting heavy pieces of artillery across difficult terrain. At the end of the event, still images of the Italian King and Queen, of poet Dante Alighieri, and finally of King George V were projected onto the screen. In May 1915 in London, a couple of weeks before Italy's declaration of war, Pettinati's Film Bureau organized a recruiting event, at the centre of which was the film *The Call Of Garibaldi* (1915). The event spanned several days and took place at two London theatres, the Sadler's Wells and the Broadway Gardens in Walham Green. Screening a patriotic film about Garibaldi at Sadler's Wells, close to Little Italy, suggests that Pettinati was trying to attract an audience of Italian immigrants. The Film Bureau arranged for the presence of several bands, buglers, boy scouts, wounded Belgian soldiers, recruiting officers and 'a good dozen of real Garibaldi Veterans [who] were greeted with hearty cheers from a huge crowd' as they proceeded through the streets to the theatre where they were given a 'tremendous' reception.[23] According to the *Kinematograph and Lantern Weekly* (*KLW*), *The Call Of Garibaldi* was 'a typical film of the Garibaldian wars, a small episode of the great feats of General Garibaldi. It is not a war film of the ordinary style, but a film full of patriotic sentiment.' The British trade paper reported that 'day after day crowds have been attracted to the two halls' and that 'no better demon-stration of the practical value of the kinema as a recruiting factor could be given'. Not only did the event result in the enlistment of 'a large number of new recruits', but the showmen were commended for their ability to 'co-operate with the Military Authorities in one of the most patriotic movements'.[24]

The May 1915 demonstration can therefore be seen as a visible result of a number of patriotic activities involving the immigrant community that began to occur months before Italy's declaration of war. Drawing on the patriotic discourse attached to Italy's unification these events aimed to cultivate a sense of national unity and belonging amongst the Italian immigrants and to give a very vigorous display of their alliance with Britain.

Arrigo Bocchi and the Pro Italia Committee

After Italy's declaration of war, Bocchi and Pettinati's activities to mobilize support amongst the Italian community increased. In late June 1915, Bocchi organized another patriotic event at the Philharmonic Hall. The event began with a lecture on the *risorgimento* delivered by Sambucetti, and, alongside the programme of films depicting the Italian army and navy, it included patriotic songs performed by Mr V. Ascoli and a performance of the first act of *Romanticismo*, Gerolamo Rovetta's play about Italy's unification.[25] Just days later, together with a group of prominent Italians and under the auspices of the Italian Ambassador and his wife, Bocchi announced the establishment of the Pro Italia Committee.

The Committee's main purpose was to raise funds in aid of the Italian Red Cross and the families of Italian soldiers and sailors in the United Kingdom. The main event in the Pro Italia Committee's 1915 calendar was the staging of an Italian Flag Day on 7 October. During the course of the day special fundraising events were held, and about 4,000 ladies, some of them dressed in Italian costume, sold little silk badges combining the national colours in flags, rosettes and flowers.[26] Chief hotels and clubs across London assisted in the Day, and restaurants decorated in green, white and red served Italian food whilst orchestras played Italian airs.[27] Although there is no mention of the screening of topical war films during the day, the film *Cabiria* (1914) was shown at the West End Cinema, accompanied by a brief speech from Sambucetti.[28] *Cabiria* had been acquired for distribution in Britain by the Tyler Film Company in April 1914, but had not been shown by mid-1915, when it was released by Horace Dickson of the Argus Film Service, accompanied by an extensive marketing campaign in the British press.[29] *Cabiria* was lauded by the trade, with the *KLW* describing it as 'a great work of art, as a great film, and as an intellectual feast in pictures … this wonderful Italian subject is of a kind to make history – kinematograph history'.[30] Dickson's re-release of *Cabiria* capitalized on Italy's recent entry into the war, and on the attention paid to the film's screenwriter, Italian poet Gabriele D'Annunzio, who had been a staunch supporter of Italy's intervention in the conflict during the period of neutrality. As well as guaranteeing a high number of ticket sales on the strength of its reputation the showing of 'D'Annunzio's Masterpiece', together with Sambucetti's patriotic speech, displayed the Italians' passionate support of the war.

Bocchi continued to play a part in the Italian immigrant community's war effort within Britain until the end of the war, by organizing fundraising activities. Italian

Flag Days occurred at least twice more on 14 December 1916 and 25 September 1918, although there is no suggestion that Bocchi held special screenings of films, Italian or otherwise, on these two occasions.[31]

Mario Pettinati and Italian war films

In 1916, Pettinati's Film Bureau began to distribute the first official Italian war films in Britain. The first of these to be seen on British screens was *On The Way To Gorizia* (*La Battaglia di Gorizia*, 1916) a three-reel film depicting the advance of the Italian troops on the city of Gorizia. The film was taken by permission of the Italian government, and although no cameraman is mentioned it seems likely that it was shot by Luca Comerio, the official Italian war cinematographer. According to the *Times*, the film gave a clear idea of the difficulties confronting the Italian Alpine troops 'in some instances 12,000 ft. above sea level amid snowstorms and blizzards'.[32]

Although the Film Bureau distributed the official war films throughout the country, they were each given special London premieres, which functioned as Italian patriotic events. On 18 September 1916, the Italian and Russian Ambassadors, Mrs Marconi, Winston Churchill MP and other prominent diplomats and personalities attended a matinee screening, arranged by Pettinati, of *On The Way To Gorizia* at the New Gallery Kinema in London's Regent Street.[33] Following the premiere, the film was given a two-month exclusive run at the cinema. During October the film was released widely throughout Britain and appears to have met with considerable popular and critical success.[34]

FIGURE 11.1 Advertisement for the war film 'On the Way to Gorizia'

A second patriotic war film screening was organized by Pettinati in November 1916. *The Battle Of The Alps (A 3000 metri sull'Adamello*, 1916) was an official record approved by the King of Italy, and sent by special courier to England for its premiere at the New Gallery Kinema on 11 November 1916.[35] The film, again shot by Comerio, depicted the Italian Alpine army's operations in the Dolomite mountain range. It showed the difficulties and dangers of transporting equipment across the mountains.

In late September 1916, Pettinati made an appeal on behalf of the Italian Red Cross Society for a nationwide Italian Cinema Day on 11 November 1916 in commemoration of the birthday of the King of Italy.[36] It is not clear how many cinemas participated in the event or whether they screened special films for the occasion, but in a letter sent to *KLW* thanking the trade paper for its support, Pettinati wrote: 'You will be pleased to hear that all the leading London kinemas have allowed either a collection or a part of the profits of their matinee, with the result that a considerable amount has been raised for the benefit of the fund.'[37] Present in the audience for the matinee of *The Battle Of The Alps* were many distinguished people including the Italian, French and Russian Ambassadors, and Queen Alexandra and the Royal Princesses, Victoria and Christian.[38] Pettinati was keen to involve not only eminent personalities but also the representatives of London-based Italian immigrant organizations and institutions in the patriotic occasion.

During 1917 Pettinati continued to display Italy's role in the war through the screening of Italian war pictures. In that year the Film Bureau's business expanded in terms of the number of fiction films the company released. These were primarily Italian productions such as *A Woman's Cavalry* (LEA, 1916) produced by the Italian firm Sabaudo Films.[39] Nonetheless, Italian war films remained an important feature of the Bureau's output, and in April 1917 he was appointed 'Official Kinema Representative of the Italian Government' in Britain, charged with promoting the work of the Italian Army to the British public through the cinema.[40] The first war film released after Pettinati's appointment was *The Battle Of The Isonzo* (1917).[41] In May, Pettinati distributed to principal cinemas throughout England a special slide displaying a photograph of the Prime Minister of Italy surmounted by the British and Italian flags. Following this, on 15 June 1917, two official Italian war films, *The Battle Of Gorizia* and *On The Heights (La Guerra d'Italia sulle Dolomiti*, 1916), were screened at the Alhambra Theatre in London. Queen Alexandra and Princesses Victoria, Mary and Maud attended the event alongside the Italian Ambassador and his wife and a number of wounded Italian soldiers. The proceeds went to the Italian Red Cross and the Italian Hospital in London.[42] Finally, in September, just prior to the unification anniversary celebrations of *Venti Settembre*, Pettinati screened the first of the latest series of official Italian war films at the Marble Arch Pavillion.[43]

Conclusion

Sambucetti and Bocchi expressed the readiness of the Italians living in London to fight alongside Britain for the common aim of freedom at the Queen's Hall months before Italy's declaration of war. The rhetoric adopted by London's Italian elite to justify and support Italy's intervention echoed the official propaganda of the Italian government, but it also acquired a special and complex meaning within the social context of Little Italy. In the run-up to Italian intervention Bocchi and Pettinati incorporated films into patriotic events targeting the Italian colony and mobilizing support for the war. Following Italy's declaration of war in May 1915, the films were positioned within events organized under the patronage of the Italian Embassy. The wider Italian colony was acknowledged on these latter occasions by the presence of representative members of the community from London-based Italian societies, Garibaldian veterans, wounded Italian soldiers, and school children. The screenings and the patriotic events in which they were situated reiterated an interventionist rhetoric based on the ideals of Italy's *risorgimento* through explicit reference to Garibaldi and Mazzini. To a certain degree war films promoted and reinforced the same patriotic impulses that guided the production of Italian historical epics. These films were perceived and deployed as instruments suited to inspire a strong Italian identity rooted in the glorious past of Rome. These idealized narratives of Italy's history conveyed a national consciousness to Italian spectators and echoed nationalist sentiments circulating in contemporary Italian political and cultural discourses. Regional and local differences were erased by an idealized notion of the 'national'. The cinematic medium was used to promote and display images of Italy as a modern and unified country, thereby reinforcing the nationalistic discourses that pervaded the Italian immigrant press. It was elite groups above all that promoted a strong sense of national affiliation and identity, which was rather different to the experiences and allegiances of the working-class members of the immigrant community.

Notes

1 'In Little Bath Street', *Manchester Guardian,* 24 May 1915, 4.
2 'The New Ally', *Times,* 28 May 1915, 4.
3 This goal was strengthened in Britain by Giuseppe Mazzini's exile to London where he founded the first school for poor Italian children. This particular connection between London's Italian community and the history of the unification of their native land formed a relevant cultural context for the reception of Italian war films.
4 B. Singer, *Melodrama and Modernity*, New York: Columbia University Press, 2001, p. 19.
5 All figures provided in this chapter are taken from my PhD thesis: 'Ethnic Audiences and Film Culture: Italian Immigrants, National Identity and the Distribution of Italian Films in London at the Beginning of the Twentieth Century', University of East Anglia, 2008.
6 See J. Buzard, 'The Grand Tour and after (1660–1840)', in P. Hulme and T. Youngs (eds) *The Cambridge Companion to Travel Writing,* Cambridge: Cambridge University Press, 2002, pp. 37–52. On the representation of Italy in the literature of the Grand Tour see C. De Seta, 'L'Italia nello specchio del Grand Tour', in C. De Seta (ed.) *Storia D'Italia, Annali 5, Il paesaggio,* Turin: Einaudi, 1982, pp. 127–263.
7 Buzard, 'The Grand Tour', in P. Hulme and T. Youngs, *The Cambridge Companion to Travel Writing,* p. 41.
8 See M. Liversidge and C. Edwards (eds) *Imagining Rome*, London: Holberton, 1996.
9 All the films I refer to in these pages have been released in the British market.
10 *The Bioscope,* 20 February 1913: 537.
11 M. Wyke, *Projecting the Past. Ancient Rome, Cinema and History*, London: Routledge, 1997, pp. 41–42.

12 G. Silvestrelli, 'La colonia italiana di Londra', *Bollettino del Ministero degli Affari Esteri*, Rome, 1895, 105–6, Italian Foreign Office Report, 1895.

13 T. Colpi, *The Italian Factor: The Italian Community in Great Britain*, Edinburgh: Mainstream, 1991, p. 55.

14 The 'very strange and foreign' procession of Our Lady of Mount Carmel was described by one observer as 'the most foreign sight that London promises in the year […] But it is in no wise a "sight" in the tourist sense […] As they passed out of sight chanting their hymn one rubbed one's eyes. All these things seen over the shoulder of Police Constable 411 on a raw English afternoon.' *Manchester Guardian,* 21 July 1902, 4.

15 'Little Italy', *Daily Graphic,* 30 November 1905.

16 'Italy in London', *Manchester Guardian,* 25 May 1915, 4.

17 *Londra-Roma,* 7 February 1914, 1. The *Società* still exists today and is now known as the Mazzini Garibaldi Club.

18 'Pictures of the Scott Expedition', *Times,* 24 January 1914, 12; 'The Wonders of India,' *Times,* 31 December 1914, 15.

19 *Kinematograph and Lantern Weekly* (hereafter *KLW* in notes and main text), 2 November 1916, 10.

20 *Kinematograph Year Book* London, 1924, 268.

21 Board of Trade 31/21175/126633 and 31/22813/140180, Public Record Office, London.

22 *Londra-Roma,* 13 February 1915, 1.

23 'Garibaldi Veterans on the Screen', *KLW,* 6 May 1915, 21.

24 Ibid.

25 *Londra-Roma,* 3 July 1915, 2.

26 *Times,* 7 October 1915, 11.

27 'Theatrical Stars as Auctioneers', *Manchester Guardian,* 3 October 1915, 3.

28 *Times,* 28 September 1915, 5.

29 The reasons for the delay are unclear, but there are several possible contributing factors. First, the length of the film in 1914 was approximately 12,000ft giving a running time of around three hours. The fact that the film had been 'cut very considerably' for its 1915 release, giving a new running time of two and a quarter hours, indicates that the *Bioscope*'s initial concerns over the film's length were warranted. See 'Cabiria', *Bioscope,* 30 April 1914, 506–10.

30 'Booming a Big Film', *KLW,* 27 May 1915, 10.

31 'Italy's Day', *Times,* 14 December 1916, 11; 'Honour to Italy', *Times,* 25 September 1918, 9.

32 'Italian War Film', *Times,* 19 September 1916, 5.

33 'Distinguished Gathering at the New Gallery Kinema', *KLW,* 21 September 1916, 28. The trade show was held on 7 September at Guido Serra's *Cines* offices in London. 'Our Gallant Ally: Italy', *KLW,* 14 September 1916, 109.

34 Advertisement for *On the Way to Gorizia* in *KLW,* 28 September 1916, xxxviii.

35 The film was also known as *Adamello.*

36 'An Italian Flag Day', *KLW,* 28 September 1916, 21.

37 'Stroller's Notes', *KLW,* 16 November 1916, 5.

38 'Queen Alexandra at the New Gallery Kinema',*KLW,* 16 November 1916, 6.

39 See, for instance, 'Film Bureau advertisement', *KLW,* 17 May 1917, 16.

40 'Italian Government Seeks the Aid of the Kinema', *KLW,* 17 May 1917, 85–88.

41 'The Battle of Isonzo', *KLW,* 12 April 1917, 16.

42 'Italian Red Cross Matinee', *Times,* 15 June 1917, 3; 'Italian War Films,' *Times,* 16 June 1917, 3.

43 'Italian Battle Pictures', *Times,* 13 September 1917, 5.

12

HOLLYWOOD IN DISGUISE

Practices of exhibition and reception of foreign films in Czechoslovakia in the 1930s

Petr Szczepanik

This chapter focuses on the ways in which foreign sound films were distributed, shown and received in Prague between 1929 and 1939.[1] Comparing the popularity of Czechoslovak, American and German productions on the local market, it presents a list of each year's top-ten hits, and draws conclusions about the short- and longer-term tendencies of local cinemagoing preferences. The chapter asks why the English language and American culture were considered to be disturbing elements by local audiences. What made German films not only more popular than American ones, but also more popular than German versions of American films? Was it the German language, which was more comprehensible to the local public than English, or the archetypes of German-Austrian popular culture represented in these films? How can we explain the extreme but short-term popularity of American talkies in the first year that they were shown in Prague, and their sharp decline in popularity in the following years? What kind of American films continued to be hits after 1930?

The basic methodological principle adopted in this chapter is to study demand rather than supply. Internationally, the works based on supply data have uniformly led to conclusions describing the homogenous and overwhelming dominance of American films on European markets.[2] The problem with such interpretations is that they conflate the quantitative primacy of supply with economic and cultural dominance. More recently, scholars have begun to investigate consumers' behaviour towards American films in Europe after the First World War. Studies by Joseph Garncarz and John Sedgwick have shown that box-office and screening data can be used to measure the economic and cultural performance of foreign film products on the local markets. According to Garncarz, who analysed charts of German film hits from 1925 to 1990, Hollywood films did not achieve greater popularity than the domestically produced films in Germany until the 1970s.[3]

The analysis presented here is based on a regionally and functionally defined sample consisting of approximately 2,500 films premiered in 20 first-run Prague movie theatres from 1929 to 1939.[4] In the study, the films were classified by the national provenance of their production companies and their popularity was estimated by the number of weeks that they were kept on each theatre's programme. An average length of Prague first-run in each year is taken to represent the estimated popularity of individual national productions. The results are presented in Table 12.2.[5] For the yearly top-ten lists presented in Table 12.6 at the end of the chapter information about the seating capacity of each theatre is given. The resulting 'popularity index' is calculated as an arithmetic product of the length of run, in weeks, and number of seats in each individual theatre.[6] Although this method reflects the actual popularity of each film more properly than the mere length of its run, it seems unnecessary to apply a seating-capacity variable for calculations of popularity by country of origin, because individual cinemas did not exclusively choose their films on the basis of their national origin. Both bigger and smaller theatres from the sample alternately exhibited Czech, American, German and French films. Because no comparable data sets are available for box-office results, for first-run releases in other Czechoslovak cities, or for subsequent runs, the procedure used here provides the only way to construct a homogenous data series that can approximately represent cinemagoing preferences in the 1930s, without exhaustively compiling data from programme advertisements in newspapers.

The structure of the Czechoslovak exhibition sector in the 1930s was determined by the fact that from the early 1920s only so-called humanitarian associations were legally entitled to hold movie theatre licences.[7] The actual business of exhibition was not undertaken by the associations, but by individual theatre owners or managers who rented the licence rights from them. Exhibitors were organized in several regional associations, but they were not integrated into chains, nor did they have any vertical ties to production or distribution.[8] The total number of movie theatres in Czechoslovakia (1,833 in 1935) was higher than in other Central European countries of comparable size: with 579,000 seats and 15.3 million inhabitants, the country had one cinema seat for every 26 inhabitants. More than 1,000 new movie houses opened during the 1920s (from 542 in 1920 to 1,513 in 1929), and despite the high costs of wiring for sound the boom continued until the total number reached 2,024 in 1932, when the Depression began to affect the national economy after a two-year delay. At the end of 1930, only Great Britain (2,500), Germany (1,864) and France (552) had more sound theatres than Czechoslovakia (148). In terms of economic performance, however, the exhibition field in Czechoslovakia was highly decentralized and composed of mostly small theatres, with 87 per cent of them not playing every day, as Table 12.1 indicates.

Taking all this into account, we can presume that the sample of approximately 20 first-run theatres in central Prague – all of them relatively spacious (average 786 seats), and wired for sound no later than the spring of 1930 – was the most viable part of the Czech market. This group of theatres had established its own association in 1928 and shared common business practices, pricing policies, advertising platforms and programming methods. While this information in itself does not provide a basis on which to claim that the group can represent the whole of the national market and its

TABLE 12.1 Theatre statistics for Central Europe in 1930

	Population (million)	Number of theatres	Population per cinema	Population per seat	Average seating capacity	Theatres with more than 500 seats		Theatres playing daily		Sound theatres (end of 1930)	
Czechoslovakia	14.7	1,817	8,107	27.4	296	180	10%	228	13%	148	8%
Germany	64.0	5,267	12,400	34.1	356	784	15%	2,106	40%	1,864	35%
Austria	6.5	869	7,500	28.4	265	78	9%	209	24%	116	13%
Poland	27.2	631	43,100	134.0	322	301	48%	200	32%	50	8%
Hungary	8.0	524	15,200	44.3	344	174	33%	220	42%	55	10%
Europe	470	33,842	13,900	34.5	405	10,878	32%	15,425	46%	5,983	18%

Source: Data for the table are taken from: J. Havelka, Čs. filmové hospodářství II. Rok 1935, Prague: Knihovna Filmového kurýru, 1936, pp. 32–7; for an overview of European cinemas in 1930 see A. Jason, Handbuch der Filmwirtschaft. Band II, Film-Europa, Berlin: Verlag für Presse, Wirtschaft und Politik, 1931, Übersicht 2 a 3 (although there are some errors in Jason's book relating to Czechoslovakia that have been corrected in the data used here). For further statistical surveys of Czechoslovak movie theatres and attendance in the 1930s, see A. Oberschall, Statistika biografů v Republice československé, Prague: Knihovna statistického obzoru, 1931; L. Pištora, 'Filmoví návštěvníci a kina na území České republiky', Iluminace, 8, 4, 1994, 35–59.

TABLE 12.2 Popularity of feature sound films according to national provenance

Number of feature sound films premiered and average length of first run

		Czechoslovakia	Germany	USA	France	German versions
1929	films	–	–	20	–	–
	weeks	–	–	4.8	–	–
1930	films	5	36	136	14	7
	weeks	9.6	3.4	2.5	2	2
1931	films	18	106	116	11	21
	weeks	6.2	2.7	2.1	1.5	2
1932	films	23	121	45	13	15
	weeks	7	2.9	2	2.8	2.4
1933	films	29	73	24	24	4
	weeks	5.1	3.9	2.7	3	0
1934	films	32	81	24	26	3
	weeks	5.1	3	2.3	3	–
1935	films	26	74	128	12	6
	weeks	4.2	2.4	2.4	2.2	3
1936	films	23	66	117	15	2
	weeks	3.8	2.3	2.6	2.4	0
1937	films	37	63	106	16	4
	weeks	3.9	1.8	2.5	2	0
1938	films	40	60	146	27	7
	weeks	3.6	1.5	2.2	2.4	0.8
1939	films	43	83	97	7	2
	weeks	4	1.7	2.5	2	–

In the cases where a number of films in a particular category is less than five, I have not enumerated the average length of their first run. The foreign versions (MLVs and dubbed) of Czech, American and French films are not included among the films from Czechoslovakia, USA and France. All dates are based on the Prague premieres, not on censorship release.

audiences, we can use other information to estimate the relationship between this sample and the whole.

A majority of the films that went on the national market in the 1930s premiered in the sample theatres. The exceptions were either commercially inferior products (particularly silent films) or German versions of American and Czech films that had been excluded from the Czech theatres in 1935 and could from then on only be exhibited in cinemas located in predominantly German territories. The performance of particular films in the sample theatres was an important indication for other cinema owners in deciding whether or not to book them. For the years 1937 and 1938 we have data identifying the sample cinemas' share of the total national box-office and attendance numbers. In 1937 the sample theatres had a total attendance of 6.17 million, or 7.3 per cent of the countrywide attendance of 84.45 million. In 1938 their attendance decreased to 5.42 million, or 7.4 per cent of the countrywide total of 72.91 million. In 1937 the sample theatres took 7 per cent of the country's total box-office revenue for Czech films (9 per cent in 1938), 15 per cent for American films

(18 per cent in 1938), 12 per cent for German films (8 per cent in 1938) and 20 per cent for French films (17 per cent in 1938).[9]

Bearing these figures in mind, it is reasonable to argue that despite the sample comprising only one per cent of the total number of theatres and being significantly more cosmopolitan and less German-oriented than the average, it is representative of the national market. Since the key thesis about Czech audiences presented in this chapter emphasizes the importance of their long-term affiliation to traditions of Austrian and German popular culture, the cosmopolitan character of the sample in fact makes my conclusions about cinemagoing preferences even more valid.

To focus my study, I have concentrated on the special case of so-called 'German versions' of non-German (mostly American) films. The German versions, widely exhibited in Prague theatres in the first half of 1930s, played an important role in public discussions about the cultural and political aspects of sound cinema, and also in negotiating the international orientation of the Czechoslovak film market. They were often considered a challenge to the dominance of the Czech language in the public sphere and implicitly to the future existence of Czech cinema as such. The German versions provide the best sample for studying differences between two key factors of the local audiences' behaviour towards foreign film products: language comprehensibility and cultural affinity.[10]

The key theoretical assumption of this study is that multiple historical rhythms distinguish the slow time of social history from the fast time of technological and political change. In making this distinction, I am applying concepts inspired by the *Annales* School's multi-temporal and multi-dimensional model of historiography to the history of moviegoing. In her book on contemporary film-historical approaches, Michèle Lagny argues that Fernand Braudel's concept of the *longue durée* is of most relevance to a discussion of reception and the so-called *mentalités* of audiences as social groups.[11] Although Braudel's concept cannot be applied to film history in its original meaning of an almost motionless 'geographical time' in which historical change is practically imperceptible, the *longue durée* is nonetheless a useful analogy through which to indicate that the collective mentalities of popular culture audiences evolve significantly more slowly than the rapidly changing 'history of events' affecting cinema through war, artistic fashion or technological change.[12] Italian cinema historian Gian Piero Brunetta has argued that different rhythms operate on different levels of cinema history: the temporality of stylistic change is not synchronous with the pace of technological innovations, shifts in critics' attitudes, an 'audience's biorhythm', or even the inertia of social life as a whole.[13] In this chapter, I try to identify some long-term cinemagoing preferences on the part of Czechoslovak audiences, which differ markedly from what traditional narrative history tells us about historical change in the development of national cinema.

The reception of German and American films

Both German and American sound films had a strong presence in Prague first-run cinemas in the 1930s. American talkies dominated the playing time of all wired theatres during the second half of 1929, but there were 36 first-runs of German sound films in 1930, 106 in 1931 and 121 in 1932. There were fewer German premieres in the next two years, as a result of new contingent regulations restricting film imports, introduced in 1932. The number of American talkies declined much more dramatically in this period, because all members of the Motion Picture Producers and Distributors of America (MPPDA) participated in a 30-month boycott of the Czechoslovak market. In late 1934 the contingent was replaced with registration fees, the Hollywood majors reached an agreement with the Czech government, and American films flooded the market again, including some older films that had been kept out of Prague cinemas by the boycott.[14]

In terms of popularity, however, the Czech sound-film market was dominated by American talkies only during the short period in the second half of 1929 when the first Czech theatres were wired for sound but Czech studios were not yet able to produce their own sound films. The first American sound films entering the Czech market achieved huge popularity as technological novelties and provoked the interest of Czech producers and distributors in talkies. The biggest hits of the season were *The Singing Fool* (1928) and *White Shadows In The South Seas* (1928), both with 12 weeks of first-run, while the most popular Czech silent film, *Pražské Švadlenky* (*Prague Seamstresses*, 1929), achieved no more than six weeks of first-run in the same year (see Table 12.6).

The reasons for this initial American success were recognized by the US commercial attaché in Prague, who reported that:

> The Czechoslovak public received them [American sound films] with enthusiasm and filled the sound cinemas despite substantially increased entrance fees. The cinema owners are thus far satisfied with the financial results – there is very little competition at present and the novelty attracts large crowds and points out that actual sound films are in great favor whereas silent films with a synchronized version are regarded less favorably, being called 'movies with phonograph accompaniment' and attract a much smaller public.[15]

By the spring of 1930 the situation had already changed. The first German talkies had arrived, some of them becoming bigger hits than the American films. The most successful of them were operettas directed by Géza von Bolváry: *Zwei Herzen im Dreiviertel-Takt* (*Two Hearts in Waltz Time*, 1930) with 17 weeks in first-run theatres, *Das Lied Ist aus* (*The Song Is Ended*, 1930) with eight weeks and *Ein Tango für Dich* (*A Tango for You*, 1930) with seven weeks.[16] In September, demonstrations by Czech fascists and other nationalist groups against German films broke out in Prague and a few other Czech cities.[17] Despite numerous chauvinistic voices in the right-wing press, even the reports from the German embassy in Prague, which were otherwise very critical of all anti-German-film tendencies, recognized that the rejection of German films did not extend to the broader public:

> American films were repeatedly screened to empty auditoria in two famous Prague movie houses … But the attendances at German sound film permit the

conclusion that a large number of Czech citizens watch German movies. The explanation lies in the fact that Czechs mostly understand, if not speak, German, while they do not know any English.[18]

In 1930 American films retained their quantitative domination and were often quite popular, but they did not achieve the spectacular successes of the previous year or of *Zwei Herzen im Dreiviertel-Takt*. On average, their first-run was almost a week shorter than that of German films (see Table 12.2). This 27-per-cent difference in popularity between German and American talkies remained constant from 1930 to 1934. Calculating the average time of the first-runs of the various non-Czech national productions (American, German, French, British) in Prague cinemas in this period demonstrates that after the relatively small number of domestically produced Czech films, German films were generally the most popular, and certainly more popular than the anti-German nationalist discourse of journalists and politicians and the anti-German demonstrations on Prague streets would suggest.

In the second half of the 1930s this situation changed to a certain extent. As Table 12.4 indicates, the popularity of American films overtook that of German films by half a week on average in the period from 1935 to 1939. However, this was less the consequence of a shift in preference on the part of local audiences than it was the result of the gradual Nazification of German film production. German film exports to most foreign countries dropped significantly in this period, although in Czechoslovakia, the popularity of German films did not significantly decline until 1937, despite their declining quality and the strong political opposition to Nazi Germany.

The popularity of American films in the second half of the 1930s shows not only a quantitative growth, but also a structural change. While the biggest Hollywood hits in the early years of the decade might have been popular because of their technical novelty, musical attractions or appealing subjects (such as films about the First World War), the results for this later period clearly demonstrate a preference for particular stars. Although many references in the period press suggest that in the mid 1920s,

TABLE 12.3 Aggregate average popularity, 1929 (1930) – 1934 (in weeks)

1. Czechoslovakia	6.6
2. Germany	3.2
3. USA	2.7
4. France	2.5
5. German versions	2.4

TABLE 12.4 Aggregate average popularity, 1935–1939 (in weeks)

1. Czechoslovakia	3.9
2. Germany	1.9
3. USA	2.4
4. France	2.2
5. German versions	0

American stars were broadly admired by both intellectuals and the wider public in Czechoslovakia, no comparable evidence is to be found in the top-ten lists from the early 1930s.[19] When the contingent regulation was lifted and Hollywood titles flooded Prague cinemas in 1935, however, Czech audiences evidently rediscovered a passion for American actors and actresses. In the three following years Czech movie-lovers' unquestionable favourite among foreign stars was Greta Garbo, whose films won top-ten positions five times (see Table 12.6).

Just a common language? Reception of German versions

Despite their dominance of supply in the 1920s, the Hollywood majors were well aware of the fact that the percentage of American feature releases in Czechoslovakia had steadily declined from 58 per cent in 1925 to 48 per cent in 1928, while German productions' share of the market grew at the same time.[20] The recommendations of the US Consulate repeatedly expressed an opinion that English dialogue could not be popular in a country where almost nobody understands it:

> With German silent films having increased at such a rate, it is only reasonable to believe that once the German talkie and sound film industry is established so that good films can be put on the market, German talking films will gradually replace American products in this market. Practically the entire population of the metropolitan area of Prague understands German, and, consequently, will be more interested in attending the showing of a film where they can understand the language than in seeing a film where unintelligible conversation is translated by captions appearing on the picture. Furthermore, in certain German sections of the country where German is spoken almost altogether, and where there are decided German sympathies, there is no doubt that the German talking films will be used almost exclusively. ... the introduction of sound and talkie films will mean, eventually, a decided falling off in American film business in Czechoslovakia.[21]

Bearing those facts in mind, American importers speculated that the best way to maintain their market position would be to shift to German versions, either dubbed or shot entirely anew on the basis of Hollywood models: the so-called multiple-language versions, or MLVs. In July 1931, the Motion Picture Division of the Bureau of Foreign and Domestic Commerce published a report by George R. Canty, Motion Picture Trade Commissioner in Europe, arguing that the only foreign films popular in Czechoslovakia were German talkies and that 'American distributors ... are reported to be doing good business only with their German dialogue product or German versions'.[22]

According to my calculations, as many as 50 sound pictures were exhibited in German versions in Prague theatres in the period from 1930 to 1934: 11 were dubbed or synchronized, two were produced only in a German version, and the rest were MLVs. The majority were produced and distributed by major American companies: Paramount (11 German versions), MGM (seven) and Fox (four), as well as six by British International Pictures.[23] From 1935 to 1937, however, the only non-German feature fiction films shown in German versions in Prague were Hungarian and Danish. The German versions of American, British and French films were still passed by the censorship board and exhibited in Czechoslovakia by the dozens, but they did not premiere in Prague first-run cinemas. This was the consequence of a special Ministry of Commerce regulation ordering that only films in 'Czechoslovak' or original versions could be imported and shown in Czechoslovak cinemas. Exceptions were permitted for regions in which more than 50 per cent of citizens were of German nationality, and after 1938 also for the only German cinema in Prague, the Urania. During the Second World War, German versions of Italian and French films flooded Prague theatres, but that was already under the very different conditions of the 'Protektorat Böhmen und Mähren'.[24]

Generally speaking, the English language could be actually heard at Czech theatres much less often than the supply data might suggest, particularly in the early sound years. Fox used both so-called international versions (with dialogue replaced by music and subtitles or intertitles) and German dubbing more extensively, while MGM and Paramount experimented with MLVs and Czech inserts. The reason behind these strategies of disguise was a legitimate fear that too much spoken English would destroy the public's interest in Hollywood products.

American films in German versions aroused great irritation on the part of film journalists and businessmen, who considered them evidence that foreign distributors treated Czechoslovakia as a German-language market. This impression was reinforced by the results of the so-called Paris agreement on sound film patents in July 1930, when Czechoslovakia became part of German electrical corporations' exclusive territories.[25] Press commentary and the statements of producers and exhibitors argued that the versions demonstrated 'an evident effort by America to make Czechoslovakia a German-speaking exploitation colony', without even attempting to produce Czech-language versions[26]:

> Nobody would object to the screening of German films produced in Berlin, but if we allow the English-talking American films to be shown in German versions in our country, we will classify ourselves as a part of German distribution sphere and we will be swamped by German versions of all American companies.[27]

The first German version shown in Prague was United Artists' *Lummox* (Herbert Brenon, 1930), which was dubbed and exhibited under the title *Der Tolpatsch*. The Czech press received it critically, focusing on the dialogue:

> Technically it is much less perfect than other American sound and talking films … and this seems to be due to this unnatural German synchronization. Those who thought German talking films might be closer to our public than American ones will be disappointed: this German is often as incomprehensible

as English for people who don't speak it. It would be different, if it was really a German film, played by German actors.[28]

In the American press, however, the Prague exhibition of *Lummox* was described as the 'first actual showing of any German-dubbed talker'. The film was supposed to test the foreign-language markets, and the results of its distribution would help the studio decide whether other 'all-talkers' would be dubbed into German.[29] In contrast to the Czech reviews, *Variety* reported that the 'Prague sensation' had received 'high praise', noting that German was widely understood by the local public.[30] From April 1930, more German-dubbed versions and then MLVs of British and American films began to appear.

With only a few exceptions, the general opinion of German versions expressed in the Czech trade press and in dailies remained unequivocally negative. This might have been caused by fears of so-called Germanization, which was thought to be a great danger for Czech culture and national unity, or else by the suspicion that the German versions were an excuse for Americans not to produce Czech versions, or more straightforwardly by the criticism that the versions were aesthetically and culturally inferior to the 'originals'. This last criticism was frequently made in reviews of films featuring non-native German speakers, such as *Casanova wider Willen* (Edward Brophy, MGM; USA, 1931 – German version of: *Parlor, Bedroom And Bath*; Edward Sedgwick, MGM; USA, 1931), where Buster Keaton was blamed for the bad German: 'Who knows, why this American comedy was imported to our country in a poor German version. Frigo [Keaton] speaks totally English German and the German actors are mostly bad.'[31] The discourse on German versions should be understood as a specific manifestation of a broader process of the national culture's adaptation to the new medium of sound film and its disturbing separation of voice and body (dubbing), or of language and its cultural milieu (MLVs).[32] Since Czech dubbing and Czech MLVs were scarce and of marginal commercial importance in the 1930s, the focus of public debate concentrated on German versions, which also served as an excuse for political struggles against a (re-)'Germanization', or a supposed American colonialism.

The *longue durée* in the history of moviegoing

Although many journalists and even the trade press usually acted as if they did not recognize the attraction of German films for Czech audiences, there are nevertheless some commentaries suggesting that beneath the surface discourse, the harmony between Czech audiences and German films was a given:

> Despite all the political sympathies and antipathies from the war and from the takeover, despite all the mass traveling to the Western-European countries and despite furious attempts to learn French and English in post-war years, we are still deeply stuck in the realm of German influences, and the German intellectual and emotional climate. One must see and hear the pleasure with which the Czech audience reacts to the film [*Zwei Herzen im Dreiviertel-Takt*], when it

understands German words and jokes without having to wait for the translation in the Czech subtitles. We have never encountered anything like this with English and French talking films, be they the most artistic masterpieces.[33]

This commentary from *České slovo* suggests that comprehensibility ('understanding words') and cultural affinity (understanding the 'intellectual and emotional climate' and jokes) were critical to the German films' success. The question arises as to which of them was the decisive factor in leading Czech audiences to watch German movies, despite their alleged anti-German nationalist attitude.

Looking at the average length of their run in Prague cinemas in Tables 12.2 and 12.3, we can conclude that 'German versions' were much less popular than German films. Between 1930 and 1932, when the presence of German versions was at its strongest in these theatres, the versions were 25 per cent less popular than German films. This number suggests that the decision of American producers and their local subsidiaries to include Czechoslovakia in the German-language market was not based on careful enough calculation, because even American talkies, which had been rapidly losing popularity since 1930, were more popular than the German versions. If we define a hit as being a film with at least eight weeks' exhibition in the first-run theatres, the only German version to qualify was *Die Fledermaus* (Karel Lamač, Vandor-Film, Paris/Ondra-Lamac-Film; Germany/France, 1931 – a version of *La Chauve-Souris*, Pierre Billon, Karel Lamač, Paris/Ondra-Lamac-Film; Germany/France, 1931). The most likely explanation for this film's success was that it had a Czech director, Karel Lamač, and featured Czech silent-film star Anny Ondra in the main role.

The significantly greater popularity of German films over German versions suggests that cultural affinity was more important for Czech audiences than language comprehensibility. The factor of comprehensibility was not strong enough to prevail over the cultural and aesthetic values ascribed to the films or over the attachment of Czech audiences to the iconography and narrative conventions of German and Austrian popular culture. The German films were popular not simply because of their language, but rather because the public ascribed special cultural values to them.

Comparing Tables 12.2, 12.3 and 12.4 with the top-ten lists in Table 12.6 reveals that the aggregate tables indicate a significantly better position for German films than

TABLE 12.5 Share of total films premiered (including all versions) to share of total entertainment tax collected in 1937, and compared to average ticket sales

	Release share		Tax share		Average ticket sales per film
	No. of films released	Percentage of total releases	(Czech. Crowns millions)	Percentage of total tax	(Czech. Crowns)
Czechoslovakia	40	14.9%	18.25m.	32.7%	2,600,000
Germany	63	23.4%	12.8m.	22.9%	900,000
USA	106	39.4%	17m.	30.7%	850,000

TABLE 12.6 Top tens in Prague movie theatres, 1929–1939

Rank	Film	Weeks	Cinemas	Popularity Index
		1929		
1	Singing Fool (USA, 1928)	12	Alfa	13,200
2	White Shadows in the South Seas (USA, 1928)	12	Kapitol	12,612
3	Four Devils (USA, 1928/29)	11	Lucerna	8,800
4	Noah's Ark (USA, 1928)	8	Adria	8,024
5	Show Boat (USA, 1929)	9	Lucerna	7,200
6	Patriot (USA, 1928)	7	Lucerna	5,600
7	The Iron Mask (USA, 1929)	5	Adria	5,015
8	Abie's Irish Rose (USA, 1928)	6	Kotva	4,800
8	Wild Orchids (USA, 1929)	6	Passage	4,722
10	Pražské švadlenky [silent] (CZ, 1929)	6	Beránek (2), Louvre (2), Světozor (2)	4,103
		1930		
1	C. a k. polní maršálek (CZ, 1930)	22	Adria (9), Fénix (13)	22,851
2	Zwei Herzen im Dreiviertel-Takt (G, 1930)	17	Passage	13,379
3	Fidlovačka (CZ, 1930)	12	Adria (6), Lucerna (6)	10,818
4	Plukovník Švec [silent] (CZ, 1929)	9	Fénix	10,800
5	All Quiet on the Western Front (USA, 1930)	9	Alfa	9,900
6	The Love Parade (USA, 1929)	12	Lucerna	9,600
7	Rio Rita (USA, 1929)	6	Alfa	6,600
8	Das Lied ist aus (G, 1930)	8	Passage	6,296
9	Westfront 1918 (G, 1930)	8	Hvězda (4), Kotva (4)	5,884
10	Když struny lkají (CZ, 1930)	5	Alfa	5,500
		1931		
1	To neznáte Hadimršku (CZ, 1931)	14	Adria (4), Fénix (6), Hvězda (4)	13,896
2	The Smiling Lieutenant (USA, 1931)	12	Kotva (6), Lucerna (6)	9,600
2	Muži v offsidu (CZ, 1931)	12	Kotva (6), Lucerna (6)	9,600
4	Třetí rota (CZ, 1931)	9	Adria (3), Fénix (3), Světozor (3)	8,784
5	Poslední bohém (CZ, 1931)	8	Fénix (4), Metro (4)	7,996
6	City Lights (USA, 1931)	7	Alfa	7,700
6	Trader Horn (USA, 1931)	7	Alfa	7,700
8	Miláčekpluku (CZ, 1931)	9	Adria (3), Hvězda (3), Světozor (3)	7,197
9	On a jeho sestra (CZ, 1931)	8	Adria (4), Hvězda (4)	6,696
10	Der Kongreß tanzt (G, 1931)	8	Kotva (4), Lucerna (4)	6,400
		1932		
1	Lelíček ve službách Sherlocka Holmesa (CZ, 1932)	17	Adria (5), Fénix (12)	19,415
2	Funebrák (CZ, 1932)	9	Alfa (9)	9,900
3	Písničkář (CZ, 1932)	12	Kotva (4), Lucerna (4), Metro (4)	9,596
4	Anton Špelec, ostrostřelec (CZ, 1932)	11	Adria (5), Hvězda (6)	9,041

TABLE 12.6 *(continued)*

Rank	Film	Weeks	Cinemas	Popularity Index
5	*Šenkýřka U divokékrásy* (CZ, 1932)	12	Adria (3), Gaumont (9)	9,039
6	*Před maturitou* (CZ, 1932)	10	Kotva (5), Lucerna (5)	8,000
7	*Kantor Ideál* (CZ, 1932)	8	Adria (3), Světozor (5)	6,634
8	*Malostranští mušketýři* (CZ, 1932)	8	Kotva (4), Lucerna (4)	6,400
8	*Quick* (G, 1932)	8	Kotva (4), Lucerna (4)	6,400
8	*Unter falscher Flagge* (G, 1932)	8	Kotva (4), Lucerna (4)	6,400
	1933			
1	*Pobočník jeho výsosti* (CZ, 1933)	15	Adria (4), Fénix (7), Hvězda (4)	15,096
2	*Madla z cihelny* (CZ, 1933)	10	Kotva (3), Lucerna (3), Metro (4)	7,996
3	*Okénko* (G, 1933)	9	Kotva (3), Lucerna (3), Metro (3)	7,197
4	*The Private Life of Henry VIII* (GB, 1933)	8	Kotva (4), Lucerna (4)	6,400
5	*Putyovka v zhizn* (USSR, 1931)	8	Fénix (3), Olympic (5)	6,095
6	*Ein Lied für dich* (G, 1933)	8	Passage (4), Světozor (4)	6,048
7	*Leise flehen meine Lieder* (A, 1933)	6	Adria	6,018
8	*Perníková chaloupka* (CZ, 1933)	8	Beránek (2), Radio (2), Roxy (2), Světozor (2)	5,846
9	*Die Nacht der grossen Liebe* (G, 1933)	7	Metro (3), Světozor (4)	5,297
10	*U svatého Antoníčka* (CZ, 1933)	8	Gaumont (3), Hollywood (3), Passage (2)	5,009
	1934			
1	*Zlatá Kateřina* (CZ, 1934)	12	Gaumont (4), Hollywood (4), Metro (4)	8,404
2	*Csibi, der Fratz* (A, 1934)	11	Gaumont (4), Passage (4), Světozor (3)	8,003
3	*Matka Kráčmerka* (CZ, 1934)	6	Adria (3), Fénix (3)	6,609
4	*Nezlobte dědečka* (CZ, 1934)	8	Adria (3), Hvězda (5)	6,364
5	*Dokudmáš maminku* (CZ, 1934)	9	Gaumont (3), Hollywood (3), Metro (3)	6,303
6	*Hrdinnýkapitán Korkorán* (CZ, 1934)	7	Adria (4), Světozor (3)	6,187
7	*Mein Herz ruft nach dir* (G, 1934)	8	Passage (4), Světozor (4)	6,048
8	*Hejrup!* (CZ, 1934)	5	Alfa	5,500
9	*Les Nuits moscovites* (F, 1934)	9	Avion (4), Juliš (5)	5,327
10	*La Bataille* (F, 1933)	8	Juliš (5), Lucerna (3)	4,897
	1935			
1	*Queen Christina* (USA, 1933)	6	Fénix (3), Lucerna (3)	5,556
1	*Mata Hari* (USA, 1931)	6	Fénix (3), Lucerna (3)	5,556
1	*Merry Widow* (USA, 1934)	6	Fénix (3), Lucerna (3)	5,556
4	*Episode* (A, 1935)	7	Passage (3.5), Světozor (3.5)	5,299
5	*Cácorka* (CZ, 1935)	7	Hollywood (2), Skaut (5)	4,899
6	*Grand Hotel* (USA, 1932)	4	Alfa	4,864

TABLE 12.6 *(continued)*

Rank	Film	Weeks	Cinemas	Popularity Index
6	*The Lives of a Bengal Lancer* (USA, 1935)	4	Alfa	4,864
6	*The Scarlet Pimpernel* (GB, 1934)	4	Alfa	4,864
9	*Hell Divers* (USA, 1931)	6	Kotva (3), Lucerna (3)	4,800
10	*Viva Villa* (USA, 1935)	5	Fénix (3), Lucerna (2)	4,756

<table>
<tr><td colspan="5" align="center">1936</td></tr>
</table>

Rank	Film	Weeks	Cinemas	Popularity Index
1	*Modern Times* (USA, 1936)	8	Alfa	9,728
2	*Rose Marie* (USA, 1936)	9	Adria (6), Avion (2), Hvězda (1)	8,269
3	*Tři muži ve sněhu* (CZ, 1936)	8	Metro (5), Světozor (3)	6,170
4	*Bonnie Scotland* (USA, 1935)	7	Avion	5,656
5	*La Kermesse héroïque* (F, 1935)	4	Alfa	4,864
6	*Jízdní hlídka* (CZ, 1936)	6	Avion (2), Lucerna (2), Metro (2)	4,814
7	*Mutiny on the Bounty* (USA, 1935)	5	Fénix (3), Lucerna (2)	4,756
8	*Anna Karenina* (USA, 1935)	5	Fénix (2.5), Lucerna (2.5)	4,630
9	*Světlo jeho očí* (CZ, 1936)	6	Flora (2), Juliš (2), Praha (2)	4,440
10	*Mr. Deeds Goes to Town* (USA, 1936)	5.5	Kotva	4,400

<table>
<tr><td colspan="5" align="center">1937</td></tr>
</table>

Rank	Film	Weeks	Cinemas	Popularity Index
1	*Svět patří nám* (CZ, 1937)	8	Fénix (4), Lucerna (4)	7,408
2	*Good Earth* (USA, 1937)	6	Fénix (3), Lucerna (3)	5,556
3	*Camille* (USA, 1936)	6	Fénix (2), Lucerna (4)	5,304
4	*Žena na rozcestí* (CZ, 1937)	6.5	Avion (2), Metro (2), Světozor (2.5)	5,214
5	*San Francisco* (USA, 1936)	5	Fénix (3), Lucerna (2)	4,756
5	*The Texas Rangers* (USA, 1936)	5	Fénix (3), Lucerna (2)	4,756
7	*Falešná kočička* (CZ, 1937)	6	Juliš (2), Metro (2), Světozor (2)	4,368
8	*Night Flight* (USA, 1933)	5	Adria (2.5), Hvězda (2.5)	4,287
9	*Bílá nemoc* (CZ, 1937)	3.5	Alfa	4,256
10	*Lidé nak ře* (CZ, 1937)	5	Adria (2), Hollywood (3)	4,172

<table>
<tr><td colspan="5" align="center">1938</td></tr>
</table>

Rank	Film	Weeks	Cinemas	Popularity Index
1	*Snow White and the Seven Dwarfs* (USA, 1937)	13	Aleš	15,808
2	*Ducháček to zařídí* (CZ, 1938)	9.5	Aleš	11,552
3	*Cech panen kutnohorských* (CZ, 1938)	8	Blaník (4), Lucerna (4)	7,484
4	*Škola základ života* (CZ, 1938)	8	Adria (4), Juliš (4)	6,292
5	*Adventures of Marco Polo* (USA, 1938)	5	Aleš	6,080
6	*Beleet parus odinokiy* (SU, 1937)	5.5	Blaník (3.5), Světozor (2)	5,282
7	*Mad about Music* (USA, 1938)	4	Aleš	4,864
8	*Svět, kde se žebrá* (CZ, 1938)	6	Lucerna (3), Pasáž (3)	4,824
9	*Bluebeard's Eighth Wife* (USA, 1938)	5	Blaník (3), Lucerna (2)	4,794
10	*Pyotr pervyi* (SU, 1937)	4.5	Blaník (2.5), Metro (2)	4,448

TABLE 12.6 *(continued)*

Rank	Film	Weeks	Cinemas	Popularity Index
		1939		
1	*Three Smart Girls Grow Up* (USA, 1939)	7	Aleš	8,512
2	*U pokladny stál* (CZ, 1939)	11	Adrie (3), Letka (4), Máj (4)	8,312
3	*Ulice zpívá* (CZ 1939)	11	Adrie (3), Letka (3), Máj (5)	8,239
4	*Cesta do hlubin študákovy duše* (CZ, 1939)	10	Adrie (3.5), Hvězda (3.5), Juliš (3)	7,758
5	*Veselá bída* (CZ, 1939)	8	Lucerna	6,552
6	*Gunga Din* (USA, 1939)	5.5	Aleš	6,688
7	*Pygmalion* (GB, 1938)	5	Aleš	6,080
8	*Suez* (USA, 1938)	6	Metro (4), Blaník (2)	5,716
9	*Cowboy and the Lady* (USA, 1938)	6	Lucerna	4,914
10	*Boys Town* (USA, 1938)	7	Juliš (1), Letka (6)	4,833

Source: The numbers of weeks and seats used in the table are adopted from statistical yearbooks written by J. Havelka, *Čs. filmové hospodářství I-VI*, Prague, 1935–1940. Seating capacity for the years 1929–1934 (not provided in the yearbooks) is taken from Havelka's statistics of cinemas for 1933: *Seznam kin v ČSR*, Praha: Čefis, 1933. In the 1935 section, a very special case had to be excluded from the top-ten list. The Czech film *Jana* (Emil Synek, Meissner; 1935) had first-runs in as many as 11 theatres. Most of them were, however, not regular first-run houses and with one exception, the film was only exhibited for one week in each venue. Although *Jana* achieved a first-run of 12 weeks, this exaggerates its real success, since 10 of the 11 cinemas decided not to prolong the programme. Its relative lack of success is also indicated in the contemporary reviews.

the top-ten lists. Table 12.6 shows that in the period from 1930 to 1939 (excluding the exceptional year of 1929) Czech films occupy 46 top-ten positions, American films 33, German nine and French three. At first sight, it is clear that American films (especially MGM productions) were much bigger hits than German ones. If, however, we look more closely at the cultural specificities of the films, we see a more complicated picture. Between the end of the 1929 boom and the implementation of the contingent system in 1932, American talkies reached the top-ten lists less often and the successful films were of a different type from the hits of 1929. The three most popular films were the musical romantic comedy *The Love Parade* (Ernst Lubitsch, 1929), the film operetta *The Smiling Lieutenant* (Ernst Lubitsch, 1931), both featuring French star Maurice Chevalier in the main role, and *All Quiet On The Western Front* (Lewis Milestone, 1930), based on Erich M. Remarque's German novel and set in a small German town and on the front line in the First World War. *Trader Horn* (1931), which had a seven-week first run, was a safari adventure in Africa that clearly fitted in with the already established fashion for Africa-themed documentaries and was part of a broader cultural phenomenon of exotic fascinations.[34] The only US production that reached top-ten ranks between 1930 and 1935 without capitalizing heavily on European or exotic themes or styles was the musical spectacle *Rio Rita* (1929). Even in the second half of the decade, when German cinema largely discredited itself in both commercial and political terms, Hollywood films never approached the popularity of *Zwei Herzen im Dreiviertel-Takt*. In general, many top-ten American films could be seen as significantly distant from any sense of American everyday life, and as 'non-American' in terms of their subject, location, style or stars. These successes not only included two other films directed by Ernst Lubitsch and five starring Greta Garbo, most of them European-styled with an international setting, but also *The Lives Of A Bengal Lancer* (1935, a drama of a regiment of British soldiers in Imperial India), *The Good Earth* (1937, a story of a farmer in China), *Night Flight* (1933, based on Antoine de Saint-Exupéry's novel), *The Adventures Of Marco Polo* (1938, with its exotic journey between Italy and China), *Mad About Music* (1938, in which a girl in a Swiss school makes up stories of her father being an explorer in Africa), *Gunga Din* (1939, a story of British soldiers in India, based on Rudyard Kipling's poem) and *Suez* (1938, a biographical account of the builder of the Suez canal).[35]

The role of American and German films in Czech theatres of the 1930s should, then, be reconsidered in terms of both their popularity and their cultural specificity. The German films were more important than has previously been recognized, not only because of their language, but more importantly because of their cultural affinities, which operated on a slower level of historical time than that of technological novelties and political changes. If we consider that the most popular Hollywood star in Prague cinemas was the Swede Greta Garbo and the most successful director the German Ernst Lubitsch, and the only western among top-ten ranks was *The Cowboy And The Lady* (1938), a comedy set in a rodeo, we must conclude that the evidence of demand tells us more about audience preferences than the bare statistics of supply.

The very great popularity of American films in 1929 was clearly a short-term phenomenon resulting from their technological novelty, a one-off sensational event functioning as a wave on the surface. More slowly moving, longer-term tendencies, in the sense suggested by the *Annales* school, persisted underneath these surface disturbances. Within a year, as competition from German and Czech sound films emerged, it was clear that the more stable hierarchy of public preferences placed Czech films first, German films second and American movies third. This hierarchy expressed the much deeper mental dispositions and cultural affiliations of the Czech audiences, rooted in the traditions of central European, Austrian and German popular culture. Those affiliations were soon to be overshadowed by other drastic events: the anti-German demonstrations that provoked a short-term ban on German films in Prague cinemas, the contingent regulations, the Nazification of the German film industry, and finally, the war that destroyed the old habits of cultural affiliation. Nevertheless, these affiliations can be traced under the surface: the average popularity of German films did not drop significantly when Hitler took power and the Nazis ousted the greatest German filmmakers and actors from their country, but only in 1937. Despite the traditional discourse affirming the existence of a 'hidden-resistance' to German occupation, the popularity of German films grew rapidly during the war (average 1.8 week first-run in 1940, 2.2 in 1941, 2.9 in 1942, 3.8 in 1943, 5 in 1944). The popularity of Czech films was also much higher during the Second World War (growing from 5.3 weeks first-run in 1940 to 9 in 1945) and the general rise in attendance is usually explained by the lack of other opportunities to spend money on entertainment and the escapist function of cinema. Even in combination with the German practices of forced releases and obligatory visits to theatres, and the presence of the German military in Prague, these factors do not completely account for the almost 200 per cent total growth of German-films' popularity during the war.[36]

For the year 1937, when Prague data show the most abrupt decrease in the popularity of German films, it is possible to compare the conclusions based on the sample with more general data which is not available for previous or subsequent years. The standard film business statistics for 1937 enumerate the amounts of entertainment tax collected across the whole country from film exhibition, and include tax figures and information on average ticket sales for films of different national origin. These statistics also provide average earnings for films from each national provenance (see Table 12.5).[37] From this countrywide data, we can conclude that German productions earned relatively more money not only in the early 1930s but throughout the decade. On that basis it is possible to hypothesize that the basic claim of the present study, that the long-term preferences of Czechoslovak audiences demonstrated their deep attachment to the traditions of German and Austrian popular culture, applies not only to the Prague first-run sample, but even more to the rest of the country and to subsequent distribution runs. On the other hand, we have to keep in mind that the nationwide numbers included vast German-speaking regions of the country.

In this context, it is symptomatic that undoubtedly the most popular figure in the whole period discussed here was the Czech comedian Vlasta Burian, whose films ranked in the top-ten lists as many as 12 times (with an average of 11.9 weeks first-run per film). Burian was bi-lingual and starred in the German versions of his Czech films, usually playing Austrian-army officers. According to his biographer, Vladimír Just,

> Burian's humour grew organically from the milieu of Prague outskirts and its pubs, cabarets and *tingeltangels* of pre-World War I provenance. In this sense, he personifies their extension and 'immortalization'; he is a projection of a certain Central European, Austro-Hungarian urban-cultural mentality in the transformed conditions of post-World War I Czechoslovakia and of the subsequent Protektorat Böhmen und Mähren.[38]

Conclusion

The social realm of *mentalités* clearly evolved at a slower historical rhythm than did changes in technology, administration, politics or international affairs. While enthusiasm for the technological novelty of sound films or demonstrations against German films surfaced and faded quickly, the deeper cultural affiliations of Czechoslovak audiences transcended the turning points of the First and Second World Wars, as well as the establishment of an independent Czechoslovakia in 1918 and its break-up into the Protektorat in 1939. The cinemagoing preferences and basic patterns of popular culture remained largely the same, because audiences remained embedded in the old cultural traditions of popular music and theatre: Viennese waltz and operetta, as mediated above all by the Viennese-style trio Géza von Bolváry (director), Robert Stolz (composer) and Willi Forst (actor) starting with their *Zwei Herzen im Dreiviertel-Takt*, the most popular foreign film of the whole decade.[39]

The Hollywood majors had to find a strategy that could overcome these cultural barriers in the local market. To get closer to the hearts of central European viewers, Hollywood took on several disguises: when the linguistic disguise provided by German MLVs and dubbing largely failed with Prague audiences and elsewhere, the best option seemed to disguise itself behind European subjects such as the First World War, European stylists such as Ernst Lubitsch and stars such as Greta Garbo, or else to resort to mythical local or exotic settings like Africa, India and China.

In more general terms, the case of cinemagoing preferences in Czechoslovakia highlights some of the intricate relations of popular culture to modernity. On one hand, it had been driven by the principles of perpetual newness, improvement, internationalization and consumer emancipation: all of the technical audio-visual media of the 1930s were based on delivering a continuous stream of novelties, be it popular-song recordings, live broadcasting or film premieres, and on promising ever richer sensual experiences via electronic gramophone replacing the mechanical, sound film replacing the silent, vacuum tube radios replacing the crystal. On the other hand, these novelties entered the same public arena, where cultural-geographic proximity and the historical persistence of mentalities played a major role, reviving stereotypes that the newly formed national public shared with the neighbouring parts of the former Austro-Hungarian Empire, and contrasting not only with the modernist tendencies of highbrow literature and theatre, but also with the official cultural politics of the Czechoslovak government, oriented toward the Little Entente, France, Great Britain, and the US.

Notes

1 I would like to thank John Sedgwick for reading this chapter and graciously providing me with useful comments. I also received valuable advice from Joseph Garncarz and Peter Krämer. This chapter elaborates on and expands propositions from an earlier essay of mine, where I discussed the reception of German films in the more limited period of 1929–34 and where I used a simpler method of constructing lists of movie hits. See P. Szczepanik, '"Tief in einem deutschen Einflussbereich": Aufführungs- und Rezeptionspraktiken deutschsprachiger Filme in der Tschechoslowakei in den frühen 1930er Jahren', in J. Distelmeyer (ed.) *Babylon in FilmEuropa: Mehrsprachen-Versionen der 1930er Jahre*, Munich: Edition Text + Kritik, 2006, pp. 89–102.

2 See Z. Štábla, 'Vývoj filmového obchodu za Rakouska-Uherska a Československé republiky (1906–39)', in I. Klimeš (ed.) *Filmový sborník historický 3*, Prague: ČFÚ, 1992, pp. 5–48; G.Heiss and I. Klimeš, 'Kulturní průmysl a politika. Československé a rakouské filmové hospodářství v politické krizi třicátých let', in G.Heiss and I. Klimeš (eds) *Obrazy času: Český a rakouský film 30. let / Bilder der Zeit: Tschechischer und österreichischer Film der 30er Jahre*, Prague: NFA, 2003, pp. 303–91. Among English-language publications, see e.g. T. H. Guback, *The International Film Industry: Western Europe and America Since 1945*, Bloomington: Indiana University Press, 1969; K. Thompson, *Exporting Entertainment: America in the World Film Market, 1907–1934*, London: BFI, 1986.

3 J. Garncarz, 'Hollywood in Germany: The Role of American Films in Germany, 1925–90', in D. W. Ellwood and R. Kroes (eds) *Hollywood in Europe: Experiences of a Cultural Hegemony*, Amsterdam: VU University Press, 1994, pp. 122–35; J. Sedgwick, *Popular Filmgoing in 1930s Britain: a Choice of Pleasures*, Exeter: Exeter University Press, 2000.

4 The number of theatres changed slightly during the period.

5 This was also the only method used to estimate popularity in period statistics and business reports, both Czech and American – see e.g. J. B. Smith, 'Imports and Production of Motion Picture Films in Czechoslovakia', *Motion Pictures Abroad*, 199, 1 February 1937, 6.

6 This 'popularity index' is inspired by John Sedgwick's POPSTAT index. The simplified version was chosen because of the required data were not all available (ticket prices for individual theatres) to strictly follow Sedgwick's methodology. See J. Sedgwick, *Popular Filmgoing in 1930s Britain: a Choice of Pleasures*, Exeter: Exeter University Press, 2000.

7 These associations were non-profit, non-governmental organizations operating in the realms of charity, sport and social service.

8 Approximately 20 theatres were unofficially controlled by the biggest production company Elekta. The only theatres that might be considered as a chain were the 800 very small cinemas licensed by the gymnastics organization Sokol, which followed common business practices.

9 See J. Havelka, *Čs. filmové hospodářství IV. Rok 1937*, Prague: Knihovna Filmového kurýru, 1938; J. Havelka, *Čs. filmové hospodářství V. Rok 1938*, Prague: Knihovna Filmového kurýru, 1939.

10 For the period audiences, the term German version referred to any replacement of an original language by another one, including multiple-language versions (MLVs), dubbing and synchronized voice-over commentary in documentaries.

11 M. Lagny, *De l'histoire du cinéma. Méthode historique et histoire du cinéma*, Paris: Armand Colin, 1992, pp. 34, 181–236.

12 Braudel regarded the history of events as the most superficial aspect of historical change, 'surface disturbances, crests of foam that the tides of history carry on their strong backs' (F. Braudel, *On History*, Chicago: The University of Chicago Press, 1980, p. 21). He encouraged the other social sciences to adopt the concept of the *longue durée* as a common methodological ground revealing the multi-temporality of their subjects and enabling comparisons between their conclusions. See F. Braudel, 'History and the Social Sciences. The *Longue Durée*', in *On History*, Chicago: The University of Chicago Press, 1980, pp. 25–54.

13 G.P. Brunetta, 'History and Historiography of Cinema', *Cinema & Cie*, 1, 2001, 98–108.

14 Detailed documentation on the three-year negotiation between American companies and the US Department of Commerce (coordinated by George Canty) on one hand and the Czechoslovak government and film-business representatives on the other, is to be found in the National Archives, Washington DC: General Records of the Department of State (Record Group 59), Decimal File, Czechoslovakia, Internal Affairs 1910–44 (860f), Microform M1218, Roll 19.

15 A. Broft, 'Motion Pictures in Czechoslovakia', *Motion Pictures Abroad*, 21, 31 March 1930, 5.

16 Filmographical data relating to the first two of Bolváry's films mentioned above are included in Table 12.6.

17 See N.M. Wingfield, 'When Film Became National: "Talkies" and the Anti-German Demonstrations of 1930 in Prague', *Austrian History Yearbook*, 29, 1, 1998, 113–38.

18 M. Alexander (ed.) *Deutsche Gesandtschaftsberichte aus Prag. Teil III*, p. 298 (manuscript).

19 This affection for American movie stars such as Mary Pickford, Douglas Fairbanks, or Harold Lloyd was commented on in the US press by Carl Laemmle during his visit to Carlsbad: see 'American Films Are Liked by the Czechoslovakians', *New York Times*, 5 September, 1926, X4.

20 J. Havelka, *Kronika našeho filmu 1898–1965*, Prague: Filmový ústav, 1967, p. 74.

21 J.W. Bailey, Jr., 'Sound Film in Czechoslovakia', *Motion Pictures Abroad*, 9, 24 October 1929, 3–4.

22 G.R. Canty, 'German Film Distribution in Foreign Countries', *Motion Pictures Abroad*, 51, 8 July 1931, 2.

23 For a detailed list of German versions of American films on the Czechoslovak market see my article 'Poněmčený Hollywood v Praze', *Iluminace*, 18, 1, 2006, 59–84.

24 In the 1920s and 1930s, data on 'nationality', i.e. on the main ethnic groups living in Czechoslovakia, were usually drawn from nationwide censuses (1921, 1930), which defined it on the basis of a maternal language. See regulation nr. 131.126/34 of Ministry

of Commerce from 14 November 1934 (with amendment from 24 January 1935, nr. 6320/35). See also J. Havelka, *Čs. filmové hospodářství I: Zvukové období 1929–1934*, Prague: Čefis, 1935, pp. 5–9; J. Havelka, *Čs. filmové hospodářství III: Rok 1936*, Prague: Knihovna Filmového kurýru, 1937, p. 19.

25 For discussion of the Paris negotiations from a German perspective, see W. Mühl-Benninghaus, *Das Ringen um den Tonfilm: Strategien der Ëlektro- und der Filmindustrie in den 20er und 30er Jahren*, Düsseldorf: Droste, 1999.

26 Troufalost Ameriky: 'Pro ČSR stačí přece německá verse!', *Filmový kurýr*, 4, 22, 30 May 1930, 2.

27 ksž [K. Smrž], 'Touha každé ženy', *Lidové noviny*, 38, 284, 6 June 1930, 11.

28 í, 'Poněmčený film v Praze', *České slovo*, 13, 15 January 1930, 5.

29 '*Lummox* Dubbed by U.A. for Publix', *Variety*, 97, 10, 12 December 1929, 5.

30 'UA's German *Lummox* Is Prague Sensation', *Variety*, 98, 2, 22 January 1930, 5; 'Dubbed *Lummox* Wins Praise in Prague Run', *Variety*, 98, 3, 29 January 1930, 4.

31 Dr O. R. [O. Rádl], 'Nové filmy v Praze. Frigo musí mluvit', *České slovo*, 258, 6 November 1931, 11.

32 For a theoretical discussion of MLVs and dubbing see N. Ďurovičová, 'Translating America: The Hollywood Multilinguals 1929–33', in R. Altman (ed.) *Sound Theory / Sound Practice*, New York – London: Routledge, 1992, pp. 139–53; N. Ďurovičová, 'Local Ghosts: Dubbing Bodies in Early Sound Cinema' in A. Antonini (ed.) *Il film e i suoi multipli/Film and Its Multiples*, Udine: Forum, 2003, pp. 83–98.

33 'O německých filmech a jejich nebezpečí kulturním', *České slovo*, 227, 25 September 1930, 5.

34 Early sound documentaries like *Am Rande der Sahara* (Martin Rikli, Rudolf Biebrach, Ufa; Germany, 1930), or *Africa Speaks!* (Walter Futter, Walter Futter Productions; USA, 1930) were tremendously successful in prime Prague theatres of the time.

35 It is probable that similar regularities in cinemagoing preferences might be detected in other European countries. It certainly applies to British audiences in the 1930s, where only Hollywood movies that were 'far removed from everyday reality' of the United States ranked among the best hits. See M. Glancy, 'The "Special Relationship" and the Cinema: Anglo-American Audiences and Film Preferences', Neale-Commonwealth Colloquium, 2005.

36 The popularity of American films cannot be considered for the Second World War period, because they were first restricted and then banned totally from Czech cinemas after the USA entered the war. For statistics and traditional resistance interpretation see J. Havelka, *Filmové hospodářství v zemích českých a na Slovensku, 1939 až 1945*, Prague: ČS. filmové nakladatelství, 1946, pp. 37, 53–58; J. Brož and M. Frída, *Historie Československého filmu v obrazech, 1930–1945*, Prague: Orbis, 1966, p. 191; L. Bartošek, *Náš film: Kapitoly z dějin (1896–1945)*, Prague: Mladá fronta, 1985, pp. 346–47.

37 The entertainment tax was derived from ticket sales of all film exhibitions except for so-called 'cultural-educational' films, and it constituted approximately 18 per cent of theatres' gross income. This tax would have been the most reliable source for estimating films' popularity, but unfortunately only overall sums for individual years are known, with exception of 1937 and partly of 1938. See J. Havelka, *Čs. filmové hospodářství IV. Rok 1937*, Prague: Knihovna Filmového kurýru, 1938, pp. 33 and 65; J. Havelka, *Čs. filmové hospodářství V. Rok 1938*, Prague: Knihovna Filmového kurýru, 1939, p. 44.

38 V. Just, *Vlasta Burian. Mystérium smíchu*, Prague: Academia, 2001, p. 186.

39 Among the other films by this trio exhibited in Prague with great success were: *Das Lied ist Aus, Ein Tango für Dich, Die Lustigen Weiber von Wien* (Super-film, Germany, 1931), *Der Raub der Mona Lisa* (Super-Film; Germany, 1931), *Ich will nicht wissen, wer du bist* (Boston-Films, Germany, 1932), *Ein Mann mit Herz* (Super-Film; Germany, 1932), *Die Nacht Der Großen Liebe* (Super-Film; Germany, 1933), etc. Viennese themes and styles were also typical for some of the number-one hits in the Second World War period, e.g. *Operette* (Willi Forst, Wien-Film; Germany, 1940) or *Wiener Blut* (Willi Forst, Deutsche Forst-Film-Produktion; Germany, 1942). Forst was later replaced by Gustav Fröhlich.

13

NEGOTIATING CINEMA'S MODERNITY

Strategies of control and audience experiences of cinema in Belgium, 1930s–1960s

Daniel Biltereyst, Philippe Meers, Kathleen Lotze and Lies Van de Vijver

Introduction

This volume has addressed the question of how cinema was conceived as both an outcome and a catalyst of modernity in various historical European settings. Its contributors have illustrated different ways of looking at cinema as a contingent part of modernization, addressing questions of cinema's social and cultural embeddedness and taking into account the everyday practices of distributing, exhibiting and consuming motion pictures, as well as their production and interpretation. In relation to the cinema-modernity debate, the collection's authors have to some extent confirmed that cinema was characterized in terms of novelty, dynamism, shock, sensationalism and entertainment; that movies were often considered to advance new or foreign (mostly American) and sometimes disturbing stories and values on issues such as femininity or consumerism; and that the act of cinemagoing was often associated with the metropolitan experience and with audiences willing to be immersed in the distracting world of cinema.[1]

At the same time, many counter-arguments have been presented that seek to draw a more ambiguous picture of this experience and record the tensions that existed between the modern and traditional values that cinema provoked. These counter-arguments have emphasized issues such as the local mediation, regulation and control exercised over the content, programming and exhibition of motion pictures. They have indicated how various social organizations, including political parties, religious groups and the state, were involved in the business of screening movies; the extent to which movies were often consumed in unexpected 'non-metropolitan' areas and places (e.g. in rural areas, in churches or opera houses), and how more generally cinemagoing was very much part of the everyday life of its audiences.

Cinema's modernity was, so to speak, far from undisputed or unmediated, and the case studies in this volume illustrate Ben Singer's conception of modernity as a 'heterogeneous arena of modern and counter-modern impulses', in which 'prominent counter-forces of anti-modern sentiment' played a significant role.[2] Following Singer's argument that 'antimodernity is understood as an essential component of the modern', this chapter concentrates on the experience and the eventual impact of these counter-forces by examining the case of Belgium in the period from the 1930s to the 1960s. The chapter is based on a series of large-scale, multi-methodological research projects on the history of film programming, exhibition and audiences' cinemagoing experiences in the northern part of Belgium, the Dutch-speaking region of Flanders.[3] In this chapter we aim to bring questions of structural control and resistance to cinema's modernity together with issues of audiences' experiences of it. Inspired by Michel de Certeau's work on everyday life, mass culture and power, the question that we address is how audiences in their practice of cinemagoing were aware of and eventually experienced top-down pressures or strategies from the forces that tried to discipline cinema.[4] After indicating how the state and various ideological-political groups tried to develop strategies by which to influence film exhibition and cinemagoing practices, we concentrate on how film viewers experienced these attempts at censoring and disciplining cinema.

Disciplining cinema in a free-market liberal democracy

From the beginnings of the film industry, Belgium was widely considered by major film producers to be a small but interesting market. One of the most industrialized and prosperous countries in Europe before the First World War, the small kingdom had a vivid film exhibition scene with a high attendance rate and a wide range of cinemas. As one of the few countries without a compulsory film censorship system for adults and with no significant film production of its own, movies from all major film production centres flew into the country, with a predominance of French and American titles.[5] In its country-by-country overview, published in 1938, the American *Film Daily Year Book* described Belgium as a quasi-unregulated and 'favoured field' of action because:

> There are no laws prohibiting foreign exchange. Money made in Belgium may be freely transferred. Certain American companies have been able by the form of their organization and presentation of appropriate accounts to avoid local fiscal levies on large sums which have been shifted to America. Local laws do not give preference to other countries over American films. There are no quota or contingent laws in effect, nor are any such laws contemplated. Legislation which might reduce or prevent American distribution of motion pictures is not at present foreseen. It is probable that Belgium will continue to be considered a favoured field by American distributors. ... There is no compulsory censorship in Belgium. When pictures are released, the distributor is not obliged by the law to submit his films to any institution for censoring them.[6]

This image of an open film market tallied with broader liberal policies in matters of politics, the economy, religion, freedom of speech and the media developed by the country's successive governments since Belgium's independence in 1830. In continental Europe, the 1831 Belgian Constitution was long considered to be a seminal piece of liberalism and popular democracy, with references to freedom of the press and an article explicitly guaranteeing that censorship could never be established.[7] These principles were also applied to other media, and when the movies became a widely popular form of entertainment and gave rise to public debates over its control, attempts to censor the medium were systematically countered on the basis of these constitutional arguments.

Free-market liberalism and media freedom, however, did not prevent laws being imposed on cinema. The first legal regulations were enacted in 1908, in the form of a Royal Decree on public safety in film venues, soon followed by a range of other regulations at city and municipality level.[8] As happened in the USA and in other Western countries, police censor boards or other local forms of content regulation were already installed before the First World War, as a result of an ever-growing protest against the dangers of cinema.[9] From 1907, the Belgian (anti-)cinema debate was fuelled on a more national level by the actions of various conservative and Catholic pressure groups. In the period preceding the German invasion of Belgium in August 1914, organizations such as the *Ligue contre l'Immoralité* (League against Immorality), the *Ligue du Cinéma Moral* (League for Moral Cinema) or the *Société belge de Pédotechnie* (Belgian Society for Paedotechnics) succeeded in making the issue of cinema as a 'school of vice' (*'l'école du vice'*), 'crime' or 'immorality' front page news in the press and prominent in the agendas of local administrations and Parliament, where the Senate debated the question in May 1914.[10] Before the war, however, no central censorship measures were taken in Belgium, although local public prosecutors and other judicial forces received orders from government to pay attention to what the influential Catholic minister of Justice, Henry Carton de Wiart, had called in 1913 'the great number of violations in these obscure theatres, most notably scenes of rape, crime and violence, as well as offences against public morality'.[11] This legislation on public order and morality became an important gateway for local burgomasters and city leaders to respond to complaints about cinemas and particular movies, a pattern of governmental behaviour that continued well into the 1970s.

During the Great War, the German occupiers applied strict censorship laws. After the war, the Belgian anti-cinema movement continued to lobby for the '*assainissement*' (literally, 'cleaning up') of cinemas. The economic and social disaster of the war helped the movement to spread its ideas about moral decay and the criminogenic effects of cinema, especially on children. These ideas were to a large extent also shared by progressive, liberal and leftist organizations, including the socialist party, which saw state initiatives on the protection of (often parentless) children as a weapon against social and economic decay. This strange alliance ensured that the issue of film censorship remained on the political agenda, and led to the September 1920 law on film control.[12] From an international perspective, the Belgian law was unique. Confronted by arguments that a film censorship law conflicted with the constitutional principle of press freedom, the Socialist minister of Justice, Emile Vandervelde, proposed a compromise solution. The 1920 law did not require film distributors or exhibitors to show their movies to a film control board unless they wanted to screen the films for audiences that included children or young adolescents under 16. Many distributors submitted films to the board for commercial reasons, while others chose to screen their movies for adults only without the board's approval. Vandervelde explicitly argued that the Belgian regulation was *not* a censorship law, that the board could not reject movies for religious, ideological or political reasons, and that it was concerned only with protecting children. From the start of its operations in 1921, however, the Belgian Board of Film Control (BeBFC) was quite severe in its every-day workings. An analysis of the board's decisions indicates that during the interwar period only one third (37.1 per cent) of the movies examined before the BeBFC received the 'children allowed' seal uncut, while another quarter (23.6 per cent) had to have material eliminated before receiving the seal. The remainder (39.2 per cent) were rejected, receiving the 'children not allowed' seal. During the 1920s and 1930s, the board was most sensitive about images of violence (39.2 per cent of all cuts), crime scenes (18.2 per cent), and depictions of sex and eroticism (22 per cent). For more than 50 years, the Belgian film control system (which is still in operation today) remained restrictive, and its common practice of cutting films was only abandoned as late as 1992.[13]

The paradox of the Belgian film control system was that while it was in principle non-obligatory and never applied to adults, it was nevertheless severe for most commercial movies seeking to reach the widest audience. On the other hand, it remained liberal in the sense that many controversial movies, including erotic movies and Soviet pictures, could be shown freely in film venues without the BeBFC's control. In some cases, however, government and diplomatic pressure was used in order to prevent particular movies being screened in the kingdom, but more important were the continuous actions undertaken by various conservative pressure groups.[14]

By far the most powerful lobby seeking to discipline the free circulation of movies was composed of a variety of Catholic organizations. As soon as cinema became a successful medium of entertainment, Catholic laymen, priests and organizations became aware of its potential influence. The Vatican and Church leaders considered this to be one of the key problems of modern society; the Belgian Father Abel Brohée, for instance, who would become one of the leaders of the international Catholic film movement, described cinema as a 'school of paganism' and 'immorality', a 'promoter of adultery', 'greed', 'evil passions', 'free love', and an 'auxiliary of socialism'.[15] Confronted by the emergence and success of commercial film theatres, particularly in rural areas where the Church still enjoyed a wide moral and social hegemony, local Catholic groups started to lobby for a 'clean' cinema and even entered into the territory of the film industry. In Belgium, as in some other Western European countries with a Catholic majority, initiatives were taken after the Great War in fields as diverse as film production, distribution, exhibition, classification and criticism.[16] By the mid-1920s, Belgian Catholics began to join forces and in February 1928 they established a network of cinemas under the umbrella of the Catholic Film Central (CFC), which aimed to operate as both an exhibition cooperative and a lobby group. In the next few years, the CFC started to classify and censor all movies on the Belgian film market for its wide range of parochial and other Catholic cinemas. Although the body it established, the Catholic Film Control Board, did not have any legal status underpinning its activities, it clearly functioned as both an additional censorship agency and an influence on the workings of the official film control board, which was persistently criticized by Catholics for being too liberal in its enforcement of cinematic morality. In 1931, the network launched Filmavox, which operated as an independent commercial distributor, working for Catholic cinemas and the CFC's members. Finally, inspired by the American Legion of Decency's successful campaign of Catholic Action, which was triumphantly followed by the Belgian Catholics, the CFC created a wider mass movement. The Catholic Film League operated through local units which played a central role in organizing Catholic-inspired screenings, controlling the morality of regular film venues, and organizing concrete actions and boycotts against 'unhealthy' pictures and cinemas.

From the end of the 1920s until the 1960s, these Catholic film organizations were considered to be among the most powerful players in nearly every segment of the film industry. The key to this power was the network of Catholic cinemas, an assortment of film venues that included many small parish halls in rural areas with only a few weekly screenings as well as large independent regular cinemas loosely associated with the CFC. The Catholic presence in the field of film exhibition certainly disturbed commercial film enterprises, but it also inspired other ideological groups to compete in the field of film exhibition. In ways quite comparable to those that Thunnis Van Oort observed for the Netherlands (in Chapter 4), Belgian society was 'vertically' divided into several segments or 'pillars' according to religion or ideology. This process of 'pillarization' involved the existence of a wide range of social institutions differentiated along these ideological lines, including political parties, schools, trade unions and hospitals.

This phenomenon of a segregated, pillarized society, which only began to disintegrate in the 1960s, also affected the field of leisure and the media, and cinema and film exhibition in particular.[17] Along with the liberal policy towards cinema at large, competition among the dominant pillars provides a key explanation as to why the statistics for the Belgian film exhibition market were usually high.[18] The main protagonists in this pillarized society were of a Catholic, socialist, Liberal and Flemish-nationalist signature, and all of them were active for some time in film exhibition as part of the propaganda and other activities by which they sought to extend their social and political influence. It is difficult to measure the power of those different blocks in the area of cinema, but a systematic in-depth investigation into the ideological signature of film venues in a selection of 57 towns with regular screening places in Flanders, presented in Table 13.1, gives some concrete indications.[19] In most of these towns or cities some form of pillarized film exhibition took place from the early 1910s onwards; in only six of them did pillarization not have an impact on the film scene. Table 13.1 shows that there was a gradual growth of cinemas until the 1950s, with the proportion of pillarized exhibition expanding from 16 per cent in the 1910s to 38 per cent three decades later. From the end of the 1950s onwards,

TABLE 13.1 Number of regular commercial and pillarized film venues in a selection of 57 Flemish towns

	Total number of cinemas	Pillarized cinemas	Catholic venues	Socialist venues	Liberal venues	Flemish-nationalist venues
1910–1920	82	13	6	4	2	1
1920–1930	149	37	16	12	7	2
1930–1940	168	56	28	14	9	5
1940–1950	167	63	38	13	9	3
1950–1960	197	61	41	12	8	
1960–1970	174	47	28	12	7	
1970–1980	99	21	13	5	3	
1980–1990	65	5	5			
1990–2000	31	1	1			

FIGURE 13.1 A special screening for clergy and other high-placed people in a film venue in Brugge in March 1935

when overall cinema attendance figures gradually crumbled, the number of pillarized film venues decreased more dramatically than commercial theatres. Among the ideologically inspired cinemas, Catholics clearly took the lead: in 38 municipalities (66 per cent of the total number surveyed) there was at least one Catholic cinema and in 13 locations Catholics were the only ones to show films on a regular basis. The socialist pillar was represented in 27 municipalities (47 per cent) and in four of them they were the only ideological group active in film exhibition. Cinemas with a Liberal signature were present in 13 municipalities, while in seven others film venues were screening movies under a Flemish-nationalist flag.

If we consider this selection of cities and municipalities to be fairly representative of the Flemish (and by extension the Belgian) film scene, we might speculate about the substantial role played for several decades by ideological film exhibition. Between the 1930s and the 1960s, between a quarter and a third of film venues were known to belong to one or another ideological or religious organization. In 1957, when the expansion of cinema venues reached its peak, a quarter of the 1,585 cinemas in Belgium would have belonged to one of the pillars. Nevertheless, one can question the extent to which these efforts to guide audiences in choosing good pictures shown in suitable venues were really effective. How successful were the pillarized theatres, and how different was their programming or atmosphere? How did people experience these cinemas, in comparison to commercial venues, or in their sense of belonging to a specific community or ideological group? How did people experience the disciplinary power behind these venues, in determining how the movies they showed were chosen and censored? These questions, which seek to examine the experience of disciplinary forces in the realm of everyday life, also apply to other powers such as state film control. Did people, for instance, know about what kind of movies were cut, and how appealing did the movies forbidden to young audiences become?

Audience tactics

These questions all relate to audiences' everyday experiences of how ideological pillar organizations, state censors and other institutions sought to influence film culture and cinemagoing habits. In considering these issues it is necessary to think about different kinds of power, and in this context Michel de Certeau's well-known distinction between strategies and tactics remains useful.[20] In *The Practice of Everyday Life*, de Certeau links the concept of strategies to a top–down power exercised by institutions acting as producers of 'will and power'. These subjects have access to a place or an institutional location that allows them to develop a strategy, order social reality and manage their relations with targets. Operating from their headquarters in Brussels, the BeBFC or the CFC could develop well-coordinated strategies intended to influence ordinary people in their cinemagoing practices. On the other hand, according to de Certeau, tactics are more often associated with individuals or consumers, who are not necessarily powerless or passive objects. Tactical power relates to everyday practices of consumption, including such activities as cinemagoing, whereby the 'weak' can 'make use of the strong' in unexpected ways according to their own use, taste and given 'opportunities'. For de Certeau these tactics, or bottom–up power, include the possibility of contestation and critique. In our case we are considering the tactics deployed by viewers to avoid the guidelines given by religious film organizations, and to manoeuvre around state controls over censored movies or those to which children were not allowed.

In an attempt to better understand such audience tactics, we have relied on the oral history components of our research projects. Our general aim was to engage with the lived experiences of ordinary moviegoers within their social, historical and cultural contexts. By means of individual in-depth interviews and inspired by cultural studies in general, as well as Annette Kuhn's 'ethnohistory' approach in particular, we wanted to investigate the role of cinema within audiences' everyday life and leisure culture.[21]

Our oral history respondents were mainly selected and found in homes for elderly people or through responses to calls in local newspapers. Although we were not looking for statistical representativeness, we strove for sufficient variation in terms of age, class, gender and ideological background. A total of 391 interviews were conducted, half of them with people living in an urban area (155 in Antwerp, 63 in Ghent), and another 173 with people from 21 smaller towns and villages in a more rural setting. The sample comprised slightly more women (52.5 per cent) than men, all born between 1912 and 1959. The length of the interviews varied according to the storytelling capacities of our respondents, with an average of around one hour per interview. The interviews dealt with particular themes, including questions in relation to the interviewees' memories of censorship, the role of the Church and other pillars.

The oral history project resulted in a set of extremely rich and diversified information about the importance of movies, the social experience of cinemagoing, and the role of this activity in everyday life.[22] Taking into account that the peak moviegoing period of people's lives was before the age of 25, the largest part of our respondents' stories focused on the period between the 1930s and mid-1970s. Although this is a very broad time span, many respondents talked about it as if it were one homogeneous period. As other oral history studies have underlined, this is partly due to the fact that memories are highly selective and subjective, distorted by time and clouded by nostalgia – all of which occlusions pose problems for interpretation. Memory is not a passive depository of facts, but an active process of creating meanings.[23] The selective workings of personal and collective memories include strategies of repetition, fragmentation, narration, the use of anecdotes, and the tactics of forgetting and of adapting past experiences to contemporary thinking about particular social issues.

These considerations have wide-ranging implications for historical research, especially for the issues at stake in this chapter where people might want to either minimize or exaggerate the past impact of state control or of the Church and other societal institutions that have lost their power today. This might particularly be the case for the Catholic Church, but also by extension for the other traditional pillars. Analysing the interview transcripts in this light, we came to the conclusion that most respondents confirmed the moral and societal power of the Roman Catholic Church for the period they were discussing. They also knew about the Catholic film initiatives because, as one respondent argued, 'in nearly every village there was a priest who had a projector ... and a film venue' (JV, male, born in 1930). According to another male respondent discussing the post-war period, the Church had effective power in the field of cinema because 'priests talked about film when they were in their pulpit' and 'yes, the church could make or break a movie' (GVV, male, 1936).

While these observations seem to confirm the Church's power, respondents often minimized its influence when discussing their own moviegoing habits, insisting with some conviction that they chose their theatres and movies freely. Ideology, it seemed, did not matter because 'people didn't worry about it':

> Even in the socialist cinema people didn't talk about politics. In fact, in none of the film venues. They did in the cafés near the cinemas, but never in there.
>
> *(AVM, male, 1919)*

People chose a cinema on the basis of the quality and attractiveness of the programme, the atmosphere, and much less on the basis of any ideological consideration:

> That was a socialist cinema, yes. But they showed all kinds of movies. American, German, you name it. The audience didn't care. The only thing that mattered was the film.
>
> *(VVS, male, 1928)*

While cinemas in a socialist network were perceived as being somewhat more adapted to the regular cinema experience, at least at the level of the variety of movies, Catholic movie venues were often seen as family cinemas, or as the ideal places to go and see movies with children and other relatives. The movies screened in these venues were chosen by the CFC or by local priests so that, as one interviewee put it, people 'knew that one shouldn't expect too many dangerous things' (GM, male, 1921). Deciding whether to visit a commercial or a Catholic theatre depended on social habits as well as on its location in the neighbourhood and the ticket price. If the purpose was to have a day out with the family and see a movie, a Catholic movie theatre became a potential choice. But when going out with their partner or with friends, Catholic theatres were a less attractive option because of their strict rules of conduct, and our Catholic respondents opted to go to a commercial theatre instead.

When we asked our Catholic respondents to describe the Catholic movie theatres in the city, it immediately became clear that this type of theatre never came close to fulfilling its purpose as an alternative to the commercial circuit. Catholic cinemas and the other venues in the pillarized system were usually cheaper, but respondents often described them as 'not real cinemas', especially when compared to city centre palaces. Many of the characteristics with which they associated Catholic movie theatres stood in direct opposition to their expectations of a 'real' cinema: Catholic venues were described as small, dark, uncomfortable and as technically less suitable for movie screenings.

> I sometimes went to Catholic cinemas. But the seats there were really uncomfortable, and that was not a real cinema. The Rex or Metro, those were cinemas. Really luxurious and with the best movies.
>
> *(EM, female, 1923)*

Programming was another point of concern. Most Catholic cinemas had only a limited number of screenings per week and often opened only at weekends. Commercial cinemas in urban areas played on a daily basis with continuous programming, so that people could walk in at any time they wanted. The Catholic venues lacked this perceived image of accessibility, which enabled cinema to become the dominant form of leisure activity. In addition, the films screened in Catholic theatres were frequently considered to be second-rate, old and inferior to the ones played at commercial theatres.

> What did they show? Those were films that were really old and totally worn-out, the leftovers really. The ones they could get at a cheap price, because they couldn't afford expensive movies. They played one or two box office hits from a few years before, but all the rest … well, that was just what they could get.
>
> *(JA, male, 1941)*

In comparison to the big cities, the tension between ideological and commercial theatres in smaller villages was much higher. In Flanders it was not uncommon for small villages to have two or three movie houses, usually including a Catholic venue and sometimes a socialist one. Many respondents indicated that in these small-scale communities with a high level of social control, local priests often preached against commercial theatres and 'bad movies'. The priests' moral power also worked on a face-to-face and interpersonal level, through school masters, parents or pressure exerted on cinema owners or other people working in the local film business:

> Our local priest was absolutely against our cinema. He regularly visited my aunt (who owned the local movie theatre) to drink coffee. But whenever he was in the pulpit during mass, he was always preaching against the cinema.
>
> *(MVD, male, 1934)*

> If our local priest heard that children had gone to the movie theatre in Merelbeke, he immediately told their parents. He had a lot of contempt for people who went to the cinema. Those people were inferior to him. He talked about that in schools.
>
> *(JD, male, 1932)*

The Catholic film organization also made sure that its film classification codes were nailed to the church door, in schools and other public spaces within the Catholic network. The codes were also published in a variety of magazines, leaflets and newspapers with a Catholic orientation. When the codes used by the CFC were mentioned to our respondents, they immediately produced a smile of recognition with almost every one of our respondents. Many Catholic respondents testified that they used the classification system exactly as intended by the Catholic leaders and that parents also used the codes to decide which movies were suitable for their children. The system, however, also had an opposite effect, with forbidden movies having a strong appeal to audiences. Stories of children sneaking into adult screenings and of Catholic adults tempted to see a forbidden movie featured prominently in our interviews.

> Sometimes I realized during a movie that my parents wouldn't have let me go and see that movie. Because of our Catholic background. And then you walked out of that movie feeling guilty. I entered the theatre out of curiosity, but I walked out knowing that I shouldn't have done that. My parents wouldn't have liked it. They wouldn't approve.
>
> *(MH, male, 1941)*

Quite similar tactics can be observed when looking at how people remembered the workings of the state control board. In general, people knew about the BeBFC's existence and tried to underline the board's effectiveness by offering dramatic examples of how movies had been cut, or by referring to the effective police control in cinemas:

> You had to be careful, because, if a cinema was caught three times on letting in children for forbidden movies, they could be closed. It was a strict control.
>
> *(MT, female, 1935)*

One of the respondents, who had worked at the Ghent police vice squad, energetically talked about different cases of prints taken into possession and other forms of

FIGURE 13.2 Cinemagoers in front of Cinema Capitole in Ghent, probably spring 1950, with the official censorship code clearly advertised for Victor Fleming's *Joan of Arc* ('children allowed' right under Bergman's name)

legal action against cinema owners. These actions attracted a lot of attention precisely because 'what is forbidden is alluring to people' (JD, male, 1935). The BeBFC's most visible strategy of disciplining or regulating cinemagoing was imposing age restrictions, a practice that was visible in the programme, on posters and movie stills. Putting stickers on the pictures, for instance 'to hide a naked breast' (GV, female, 1937), was a common practice, but again it often had the opposite effect from that intended, and for some *risqué* cinemas amounted to a marketing strategy. Many respondents joyfully recalled adventurous attempts to get into cinemas where movies with age restrictions were scheduled. The interviews revealed an array of ingenious tactics, from using makeup and dressing like an adult ('we dressed like our mothers, a long skirt and blouse, a *décolleté*, and so we could enter' (AF, female, 1942)) to manipulating identity cards:

> My friends could get in but they were all one year older than me and I had an identity card and that was written with a pen. I have one indicating 1941. I made it a zero and then I could get in.
>
> *(HGT, male, 1941)*

These bold stories about evading controls at the cinema's entrance are, however, put into a less audacious perspective if we take into account the numerous stories about the weak enforcement of age restrictions, and the many descriptions of how easy it was to get in. As a former policeman said,

> There were many cinemas in the city. We went to the cinema about once a week. And then we were looking for minors and that's about it. Often we went to a cinema, just for ourselves, to watch a movie or something.
>
> *(JDM, male, 1944)*

Another BeBFC strategy which was not publicly announced was cutting movies. The interviews, however, showed that people knew about this and complained about it, although not necessarily with great intensity. One exhibitor suggested that

> Sometimes there was a little controversy around it. But I do not remember that people, so to say, were traumatized by it [laughs].
>
> *(JD, male, 1942)*

People were most often annoyed about cuts because they limited the pleasure of watching a movie:

> Yeah, those pieces cut out of the movie were really annoying, and you could see it technically: they were badly put together. They were technically bad. Usually it was a nude scene … Even censorship of language, like swearing for instance. And then there was a 'beep'. It was terrible. Ah, we often talked about it.
>
> *(AL, female, 1942)*

Technical problems of clumsy editing and sound were only one issue. People were most aware of censorship through its effect on the film's narration, with scenes of violence and eroticism often being deleted. These cuts were usually experienced as annoying cosmetic exclusions, but in some cases deletions caused real problems to an audience's understanding of the story's continuity. Too many cuts could provoke an audience response, as in the case of the juvenile delinquency movie *Rock Around the Clock*, where many cuts drove the young viewers to 'yell and jump on their seats' (MT, female, 1935). Another extreme form of rebelling against the BeBFC's strategy was to watch the movie elsewhere:

> If we wanted to see that movie, we went to France ... to Lille ... half an hour by car ... In France they showed more. Yes, we often drove to Lille because my husband said, they give the original version there, no cuts.
>
> *(AL, female, 1942)*

These were extreme reactions, and people more frequently argued that cuts were, so to speak, part of the game. To some extent, people seemed to accept and downplay those practices, although they now often denounce them nostalgically as obsolete and paternalist:

> You could see it, because then there was a problem with the story. You knew that there was a jump in it. But the pieces that were cut out ... these were usually no atrocities, but at that time a bare knee was already too much.
>
> *(CW, male, 1938)*

Again, the productivity of the censorship system often lay in the attractiveness of getting around the system, or in being attracted to the forbidden fruit[24]:

> Sure, as a Catholic you had to be good. But these quotations in the newspapers were excellent to show you where kids weren't allowed, so you'd know that those were the ones you definitely had to see!
>
> *(AV, male, 1933)*

Conclusion

The Belgian case study illustrates the many paradoxes and ambiguities that Singer referred to in discussing modernity and cinema's role in it. As a result of the liberal film policy, cinema was a booming business in Belgium, and many different types of film exploitation emerged, both in metropolitan and small-town, as well as in rural areas. Cinema was not only a commercial enterprise, but also attracted many unexpected players, including organizations that were at the heart of the ideologically strongly segregated Belgian society. As well as being made for entertainment, films were also part of the politicized attempts to attract people into a particular pillar's sphere of influence, or to keep them within it. In Belgium Catholic leaders, who had long followed the Pope in his view that cinema was a dangerous 'modern means of diversion', developed an integrated strategy which in turn attracted the Vatican's attention when it reformulated its view on the 'film problem'.[25]

Although Belgium's political and business elites wanted to present the country as an example of free trade, free-thinking and freedom of expression, they could not resist the campaign to discipline the film medium. The state control system, which was in theory more liberal than in most other Western countries, in practice soon became a restrictive instrument with which to censor the movies. Not only was film content regulated by the state, but Catholic film organizations also built a widely publicized film classification system. This situation resulted in two separate systems of censorship and classification codes, and in various forms of control–informal, social control on their compliance as well as official police and judicial control.

In this chapter we considered the counter-forces' strategies as an inherent part of modernity, exemplifying Singer's concept of 'ambimodernity'. In an attempt to assess the effectiveness of these strategies of control on audiences and their cinemagoing habits, we applied methods coming from recent historical audience and oral history research approaches. We do not want to provide definitive answers on issues such as the impact of censorship or the possible forms of resistance, but we do come to the conclusion that people were very much aware of these forces and frequently tried to evade or escape them, up to the point where systems of control could have an inverse effect by promoting censored movies. Following de Certeau's terminology, we have located these audience tactics in attempts to change identities by, for example, trying to look older, and in attempts to circumvent authorities' control or to be attentive to what was forbidden. In the respondents' memories, however, these tactics frequently remained innocent and playful, at least in the context of cinema, rather than con-stituting a conscious act of resistance. Film censorship or the Catholics' attempt to create a pillarized viewing pattern were explicitly denounced by some, and it was often seen as annoying, but people also seemed to look on it as part of the game of viewing, and as acceptable as long as people believed they had a choice.

Overall, this was neither a clear-cut top-down, nor a fully bottom-up story. Experiences of control and censorship in cinema thus illuminate the classical trope of structure versus agency. The tension between strategies of control and the everyday life tactics of audiences makes a strong case for a (new) cinema history whereby lived realities add a substantial layer to institutional histories.

Notes

1 See Ben Singer's discussion in Chapter 4 of *Melodrama and Modernity: Early Sensational Cinema and its Contexts,* New York: Columbia University Press, 2001. See also L. Charney and V.R. Schwartz (eds) *Cinema and the Invention of Modern Life*, Berkeley, Los Angeles, London: University of California Press, 1995; C. Keil and S. Stamp (eds) *American Cinema's Transitional Era*, Berkeley: University of California Press, 2004; F. Cassetti, *Film, Experience, Modernity*, New York: Columbia University Press, 2005.

2 See B. Singer, 'The Ambimodernity of Early Cinema: Problems and Paradoxes in the Film-and-Modernity Discourse', in A. Ligensa and K. Kreimeier (eds) *Film 1900: Technology, Perception, Culture,* New Barnet: John Libbey, 2009, p. 38.

3 The Kingdom of Belgium covers an area of 30,528 square km, and it has a population of about 10.8 million people. It consists of three parts: a French-speaking southern region (Wallonia), a densely populated, Dutch-speaking region in the northern part (Flanders), and the bilingual capital of Brussels, which is composed of 19 municipalities and over 1,000,000 inhabitants. In 2011 Flanders had a surface area of 13,522 square km, a population of 5.9 million and a population density of 434 inhabitants per square km. Flanders is divided into five provinces and 308 municipalities. Two of them are metropolitan cities (Antwerp and Ghent). The data in this article are based on three research projects: (1) a joint University of Antwerp and Ghent University research project: *'The Enlightened City': Screen culture between ideology, economics and experience. A study of the social role of film exhibition and film consumption in Flanders (1895–2004) in interaction with modernity and urbanization* (Scientific Research Fund Flanders/FWO-Vlaanderen, 2005–8); (2) *Gent Kinemastad. A multimethodological research project on the history of film exhibition, programming and cinemagoing in Ghent and its suburbs (1896–2010) as a case within a comparative New Cinema History perspective* (Ghent U Research Council BOF, 2009–12); and (3) *Antwerpen Kinemastad. A media historic research on the post-war development of film exhibition and reception in Antwerp (1945–1995) with a special focus on the Rex concern* (Antwerp U Research Council BOF, 2009–11).

4 M. de Certeau, *The Practice of Everyday Life*, Berkeley: University of California Press, 1984.

5 On the Belgian case as exceptional, see N. March Hunnings, *Film Censors and the Law*, Liverpool: Allen & Unwin, 1967, pp. 394–95. G. Phelps, *Film Censorship*, London: Gollancz, 1975, p. 242.

6 'Belgium', in J. Alicoate (ed.) *The 1938 Film Daily Year Book*, New York: The Film Daily, 1938, p. 1173.

7 K. Rimanque, *De Grondwet toegelicht, gewikt en gewogen*, Antwerpen: Intersentia, 2005. See also J.A. Hagwood, 'Liberalism and constitutional developments', in J. Bury (ed.) *The Cambridge Modern History: The Zenith of European Power 1830–1870*, London: CUP, 1960, pp. 190–92.

8 Royal Decree, 13 July 1908. See G. Convents, 'Ontstaan en vroege ontwikkeling van het Vlaamse bioscoopwezen (1905/1908–14)', in D. Biltereyst and Ph. Meers (eds) *De Verlichte Stad*, Leuven: LannooCampus, 2007, p. 31.

9 L. Depauw and D. Biltereyst, 'De kruistocht tegen de slechte cinema. Over de aanloop en de start van de Belgische filmkeuring (1911 – 1929)', *Tijdschrift voor Mediageschiedenis* 8, 2005: 3–26. On the USA, see L. Grieveson, *Policing Cinema: Movies and Censorship in Early-Twentieth-Century America*, Berkeley: University of California Press, 2004, p. 23.

10 P. Wets, *La Guerre et l'Enfant*, Mol: Ecole de Bienfaisance de l'Etat, 1919. A. Collette, *Moralité et immoralité au cinéma*, Liège: Université de Liège, 1993 (MA Thesis).

11 V. Plas, *L'enfant et le Cinéma*, Anderlecht: Cops, 1914, pp. 48–49.

12 Depauw and Biltereyst, 'De kruistocht tegen de slechte cinema'.

13 These data are based on the results from the research project *Forbidden Images* (FWO, 2003–6). See D. Biltereyst, L. Depauw and L. Desmet, *Forbidden Images A longitudinal research project on the history of the Belgian Board of Film Classification (1920–2003)*, Gent: Academia Press, 2008. For an analysis of crime and violence, see L. Depauw, 'Paniek in Context. Een interdisciplinair, multimethodisch onderzoek naar het publieke debat over geweld in film tijdens het Interbellum in België', Ghent: Department of Communication Sciences, 2009 (Unpublished PhD thesis).

14 See, for instance, a case study on Herbert Wilcox's controversial Edith Cavell movie *Dawn* (1928): D. Biltereyst and L. Depauw, 'Internationale diplomatie, film en de zaak *Dawn*. Over de historische receptie van en de diplomatieke problemen rond de film *Dawn* (1928) in België', *Belgisch Tijdschrift voor Nieuwste Geschiedenis/Revue Belge d'Histoire Contemporaine* 36, 2006: 127–55.

15 A. Brohée, *Les Catholiques et le Problème du Cinéma*, Louvain: Sécrétariat Général d'Action Catholique, 1927, pp. 5–13.

16 See for instance R. Molhant, *Les Catholiques et le Cinéma: Une étrange histoire de craintes et de passions. Les débuts: 1895–1935,* Brussels: OCIC, 2000. T. Trumpbour, *Selling Hollywood to the World,* Cambridge, Cambridge University Press, 2001. On Belgium, see D. Biltereyst, 'The Roman Catholic Church and Film Exhibition in Belgium, 1926–40', *Historical Journal of Film, Radio and Television* 27, 2007: 193–214. On Italy, D. Treveri-Gennari, *Post-war Italian Cinema: American Interests, Vatican Interests,* New York: Routledge, 2009. On the USA, see F. Walsh, *Sin and Censorship: The Catholic Church and the Motion Picture Industry,* New Haven, CT: Yale University Press, 1996.

17 See J. Billiet (ed.) *Tussen bescherming en verovering. Sociologen en historici over verzuiling,* Leuven: Universitaire Pers, 1988.

18 See data in D. Biltereyst and Ph. Meers, *De Verlichte Stad,* appendices. For an international comparison, see K. Dibbets, 'Het taboe van de Nederlandse filmcultuur: Neutraal in een verzuild land', *Tijdschrift voor Mediageschiedenis* 9, 2006: 46. See also J. Boter, C. Pafort-Overduin and J. Sedgwick, *The impact of the Protestant Ethic on cinemagoing in the Netherlands in the mid-1930s: an analytical comparison with the English speaking world,* unpublished paper, 2010. C. Pafort-Overduin, J. Sedgwick and J. Boter, 'Drie mogelijke verklaringen voor de stagnerende ontwikkeling van het Nederlandse bioscoopbedrijf in de jaren dertig', *Tijdschrift voor Mediageschiedenis* 13, 2010: 37–59.

19 The selection of case studies was quite representative for the Flemish region (sample of 57 out of a total of 308 municipalities and cities, or 19 per cent). The case studies primarily relate to urban and small urban areas, with a few representatives from rural areas. Some of the major cities are Aalst, Ieper, Kortrijk, Lier and Ostend. We did not select major metropolitan areas like Antwerp, Brussels and Gent.

20 M. de Certeau, *The Practice of Everyday Life,* pp. xvii, 35–37.

21 A. Kuhn, *An Everyday Magic. Cinema and Cultural Memory,* London: I.B.Tauris, 2002, p. 6.

22 For more information on the oral history see: Ph. Meers, D. Biltereyst and L. Van de Vijver, 'Metropolitan vs rural cinemagoing in Flanders, 1925–75', *Screen* 5, 2010: 272–80.

23 A. Portelli, 'What makes oral history different', *The Oral History Reader,* London: Routledge, 1998, p. 66.

24 See on this concept in relation to film censorship, A. Kuhn, *Cinema, censorship, and sexuality, 1909–1925,* London: Routledge, 1988. B. Müller (ed.) *Censorship & Cultural Regulation in the Modern Age,* Amsterdam: Rodopi, 2004.

25 Quote from Pope Pius XI's encyclical letter *Vigilanti Cura* (Vatican, 1936). See: www.vatican.va/holy_father/pius_xi/encyclicals/documents/hf_p-xi_enc_29061936_vigilanti-cura_it.html. See also Trumpbour, *Selling Hollywood to the World,* p. 213. Biltereyst, 'The Roman Catholic Church and Film Exhibition in Belgium'.

INDEX

700   500
800   300
       600
       350
        60
       700
       100
      ─────
      2610

10.